Quest for a Phantom Strait

The *Belgica* anchored in the Neumayer Channel after the discovery of the Gerlache Strait in 1898. Mt William on Anvers Island towers above the ship

Quest for a Phantom Strait

The Saga of the Pioneer Antarctic
Peninsula Expeditions 1897–1905

DAVID E. YELVERTON

POLAR
PUBLISHING

© David E. Yelverton, 2004

All rights reserved. No part of this publication may be reproduced, stored in a retrieval system or transmitted in any form or by any means, electronic, mechanical, photocopying, recording or otherwise without the permission of the publisher.

Published by
Polar Publishing Limited
10 Meadow Road
Burpham
Guildford
Surrey GU4 7LW

David E. Yelverton has asserted his right under the Copyright, Designs and Patents Act 1998 to be identified as the author of this work.

Maps on pages 10, 17, 23, 43, 54 produced by the author.

British Library Cataloguing in Publication Data
Yelverton, David E.
  Quest for a phantom strait : the saga of the pioneer Antarctic Peninsula expeditions, 1897–1905
  1. Belgian Antarctic Expedition : 1898–1899 2. Swedish Antarctic Expedition : 1901–1903 3. French Antarctic Expedition : 1903–1905 4. Antarctica – Discovery and exploration – Belgian 5. Antarctica – Discovery and exploration – Swedish 6. Antarctica – Discovery and exploration – French
  I. Title
  919.8'9'04

ISBN 0–9548003–0–3

Page design and typesetting by Wileman Design, Farnham, Surrey
Printed and bound in Great Britain by Cambridge University Press

Contents

List of Illustrations and Maps	vi
Foreword by Sir Ranulph Fiennes	viii
Acknowledgements	ix
Introduction	xi
1 The Belgian Antarctic Expedition 1897–1899	1
2 The Swedish Antarctic Expedition 1901–1903	19
3 The French Antarctic Expedition 1903–1905	39
Appendix: Historical Notes and Lists of Expedition Members	59

List of Illustrations and Maps

Illustrations

Front cover: The *Français* at Port Lockroy	
Frontispiece: The *Belgica* anchored in the Neumayer Channel after the discovery of the Gerlache Strait in 1898. Mt William on Anvers Island towers above the ship	ii
Moonlit photo of the *Belgica* trapped in the pack-ice in the winter of 1898	1
Lt Adrien de Gerlache de Gomery, Commandant of the Expedition	2
Dr Frederick Cook, Surgeon	2
Roald Amundsen, Second Officer	2
Lt Georges Lecointe, Second-in-command and First Officer	2
The wardroom of the *Belgica*: (left to right) Arctowski, Lecointe, Racovitza, de Gerlache	12
Cook's tent photographed on the abortive sledge journey on 30 July 1898. The asymetrical shape was vulnerable to strong wind on the entry side	14
The shore party on board the *Antarctic*: (left to right) Bodman, Akerlündh, Nordenskjöld, Jonassen, Sobral, Ekelöf	19
The southern party ready to start: (left to right) Nordenskjöld, Sobral, Jonassen. Note the absence of skis and the Falklands sheepdog behind the four Greenland dogs	20
Hope Bay: X indicates the position of the stone shelter and F indicates the place where the fossils were found	32
Mesozoic fossil leaves found by Dr Andersson on 1 January 1903 *Sphenopteris Nauckhoffiana* (no. 26) was the most crucial in dating the plants as having flourished in the Cretaceous period, some 130 million years ago	33
Abandon ship! Larsen's dramatic photo of the fatally-holed *Antarctic* as the crew abandoned ship	35

The return of the *Uruguay*. Seen here being towed to her berth in Buenos Aires with fore and main topmasts missing, mute witness among the jubilations surrounding the second escape of the rescued men	37
Jean Charcot aboard the *Français*. Picture by courtesy of his late daughter and her husband M. Robert Allart-Charcot	39
L'Etat-Major (officers and scientists) of the *Français*: (front, left to right) J. Rey, J-B. Charcot, A. Matha; (back, left to right) P. Pléneau, J. Turquet, E. Gourdon	40
Charcot's *Français* under sail in the pack-ice and (below) anchored at Port Charcot before wintering there in 1904	46
Some of the crew beside the hut they erected onshore for emergency stores and shelter in case the ship was crushed by the pack-ice	46

Maps

Part of L. Friederichsen's 1895 map of Carl A. Larsen's 1893 discoveries, showing the supposed Bismarck Strait and appearing to confirm the existence of Admiral Dumont d'Urville's supposed 'Orléans Channel'	xiv
The discovery of the Gerlache Strait and Lemaire Channel showing Lecointe's course in the *Belgica* while de Gerlache's party were ashore on Brabant Island and (right) Lecointe's detail of the turn away north-west to the 'Iles Cruls' on 13 February 1898	10
The track of the *Belgica* from 13 December 1897 to 28 March 1899; from charts by Lt Georges Lecointe published with the scientific reports	17
The Swedish Antarctic Expedition 1901–1903: tracks of the *Antarctic* to Snow Hill and Nordenskjöld's sledge journeys	23
The French Antarctic Expedition 1903–1905: track of the *Français* in 1904, showing coastline discovered; traced from map published with the official expedition journal showing Lt Matha's co-ordinates with longitude converted to degrees west of Greenwich	43
The French Antarctic Expedition 1903–1905: track of the *Français* in January 1905	54

Foreword

In 1895 the world's leading geographers met in London to debate the last great question of terrestrial exploration. What exactly lay at the bottom of the world? A seventh continent, or just a group of icy isles in a frozen sea?

This excellent and meticulously researched book by polar historian David Yelverton records the achievements and harrowing experiences of three of the six expeditions that grew out of that 1895 London meeting. Privately mounted, they revealed much of the Antarctic Peninsula we know today.

Led off by the Belgians in 1897, their polar apprenticeship was replete with every possible horror from scurvy to death and madness as their ship lay trapped in the ice for a year. The Swedish and French expeditions that followed them as the twentieth century dawned faced the perils of shipwreck that marooned the Swedes, and was only narrowly avoided by the French.

In parallel with Scott's first penetration of the Ross Ice Shelf, the heroic strivings of these less widely remembered expeditions deserve an honoured place in the annals of the so-called Heroic Age of Antarctic exploration which led to the great journeys of Shackleton, Amundsen and Scott that conquered the South Pole and began to unlock the mysteries of Antarctica.

This is a highly readable account of a little known, but critical, part of our polar history.

RANULPH FIENNES
Exmoor, 2004

Acknowledgements

My thanks go foremostly to the late Sir George Deacon, founder of the Institute of Oceanographic Sciences, for his generous and unstinting help in aiding a then newcomer to Antarctic history whose eyes were opened, during researches at that institute in the early 1980s, to the remarkable dramas of the pioneer Antarctic Peninsula expeditions. To Pauline Simpson, now head of information services (among other functions) at the Southampton Oceanography Centre (UK), must go a generous measure of thanks for her help at that time and on a recent visit.

Likewise, I am grateful for much guidance as to sources from Robert K. Headland, curator and archivist at the Scott Polar Research Institute, and a great deal of help from Shirley Sawtell and colleagues in that institute's world-famous polar library.

Similarly, to the descendants of the leaders and members of the expeditions I owe much gratitude for permission to reproduce photographs that were published in the accounts of the expeditions by their illustrious forebears, as also to the Library of Congress, Prints and Photographs Division and the Cook Society for permission to reproduce photos taken by Dr Frederick A. Cook on the Belgian expedition.

My thanks also go to the late Monsieur and Madame Robert and Monique Allart-Charcot (she being Commandant Charcot's youngest daughter) for generous encouragement and permission to copy the photo of her father aboard the *Français* at the explorer's home in the Rue St James in Paris, where the commandant's study is maintained exactly as it was in his day. Encouragement from Charcot's biographer, the late Mlle Marthe Emmanuel, was also very greatly appreciated.

To my publisher and editor, Solveig Gardner Servian, also goes much appreciation of her patient work in moulding my manuscript into shape and working all hours to ensure that these stories were published in time to reach the ports of call for the 2004/5 Antarctic tourist season.

Accounts of the expeditions related here are derived from: *Quinze Mois dans l'Antarctique* by Cdt Adrien de Gerlache de Gomery (Brussels, Bulens, 1902); *Aux Pays des Manchots* by Lt Georges Lecointe (Brussels, Schepens, 1904);

Through the First Antarctic Night by Dr Frederick A. Cook (London, Heinemann, 1900); *Antarctica: Två Ar Bland Sydpolens* by Dr Otto Nordenskjöld (Stockholm, Bonniers, 1904) (in particular, chapters by Dr Johan Gunnar Andersson and Captain Carl A. Larsen that were omitted from the English edition *Antarctica or Two Years amongst the Ice of the South Pole*, London, Hurst & Blackett, 1905); and *Le Français au Pole Sud* by Cdt Jean Baptiste Charcot (Paris, Flammarion, 1906).

In Chapter 1, quoted statements by Carsten Borchgrevink and Lt Georges Lecointe are respectively derived from the *Report of the Sixth International Geographical Congress held in London 1895* (London, John Murray, 1896: p. 174) and the latter's account of the Belgian expedition mentioned above. In Chapter 3, Sir Clements Markham's quoted words 'little chance' are from his paper on the projected British National Antarctic Expedition (1901–1904) in the London Royal Geographical Society's Antarctic Archive of papers concerning that expedition (ref: AA1/10/3: p. 26).

The following sources are acknowledged for illustrations:
Cover: Charcot, J.B. (op. cit.), courtesy Scott Polar Research Institute
Frontispiece: Cook, F.A. (op. cit.)
Chapter 1:
Moonlit photo of the *Belgica* trapped in the pack-ice: Cook, F.A. (op. cit.)
Portrait of Lt de Gerlache: Mill, H.R., *Siege of the South Pole* (London, Alston Rivers, 1905)
Portraits of Roald Amundsen and Dr Cook: Cook, F.A. (op. cit.)
Portrait of Lt Lecointe: Lecointe, G. (op. cit.)
Wardroom of the *Belgica*: Cook, F.A. (op.cit.)
Cook's tent: Lecointe, G. (op. cit.)
Chapter 2:
Shore party on board the *Antarctic*: Duse, S.A., *Bland Pingviner och Sälar* (Stockholm, Beijers, 1905)
Southern party ready to start: Nordenskjöld, O. (op. cit., Vol. 1)
Hope Bay: Nordenskjöld, O. (op. cit., Vol. 2)
Mesozoic fossil leaves: Halle, T.G., *The Mesozoic Flora of Graham Land* in expedition scientific reports, band 3, leif 14 (Stockholm, Lithografische Institut der Generalstabs, 1916)
Abandon ship!: Nordenskjöld, O. (op. cit., Vol. 2)
Return of the *Uruguay*: Anon, *La Argentina en los Mares Antarticos* (Buenos Aires, Ortega & Radaelli, 1903)
Chapter 3:
Charcot aboard the *Français*: courtesy of the late M. Robert and Mme Monique Allart-Charcot
L'Etat-Major: Charcot, J.B., Introduction to scientific report *Hydrographie, Meteorologie, Magnetisme* (Paris, Masson, 1908)
Français in the pack-ice; *Français* in Port Charcot; and crew beside hut at Port Charcot: Charcot, J.B. (op. cit.)

INTRODUCTION

Just Islands, or Part of an Undiscovered Continent?

In the shadow of simmering German and Boer resentments, amid the menace of accelerating naval rearmament, the western world's leading geographers met in London in 1895 to debate the most elusive challenge still facing the world of exploration: did a seventh continent exist at the South Pole? Or was there just a polar ocean with a scattering of islands, among which Sir James Clark Ross's South Victoria Land, discovered almost 60 years earlier, might be a southern Greenland?

For some 50 years Sir Clements Markham, by then president of the Royal Geographical Society in London, and the German oceanographer, Privy Councillor Georg von Neumayer, had campaigned to bring the quest to the top of the geographical and scientific agenda. At the London congress they were at last rewarded with a resolution that scientific societies the world over should urge that exploration to settle the question be undertaken before the end of the century.

As they well knew, the key to achieving that was funding for a great international multi-expedition assault on the unknown South, and that required money of an order that only western national exchequers could provide. It being hardly the opportune moment for politicians to welcome such extraneous demands, it would take the principal protagonists four years (slightly less in the German case) to travel the frustrating road to that goal.

As so often, private enterprise was ahead of establishment action. While the congress was meeting, a Belgian naval lieutenant, Adrien de Gerlache de Gomery, on special leave aboard the Norwegian sealer *Castor* in Arctic waters, was intent on gaining the experience necessary to lead a private Antarctic expedition, for which he had just won the backing of a leading Belgian industrialist.

When de Gerlache returned to Brussels in the autumn, he gained the backing of the Royal Belgian Geographical Society with the help of the congress resolution. Consequently, in the summer of 1897 his expedition sailed south, headed for the Weddell Sea via the supposed 'Canal d'Orléans', separating Louis Philippe Land from Trinity Island, reported by the French Admiral Dumont d'Urville in 1838 and apparently confirmed in 1893 by the Norwegian Carl Anton Larsen.

Larsen had sailed south for the Oceana Steamship Company of Hamburg, in command of the *Jason*, and in company with the *Castor* (Captain Pedersen) and the *Hertha* (Captain Evensen), on a whaling expedition. While the other ships headed for the Bellingshausen Sea on the west side of Biscoe's Graham Land, Larsen took the Weddell Sea route, and in an unusually favourable season, sighted its western coasts as far south as 78° 10' S. On the way he had skied to the extinct volcano Mt Christensen on the north side of Robertson Island, which he named after the captain of the Dundee whaler *Active* (later to command the *Scotia* for William Bruce) which he had encountered during the voyage. From there he could see no land to the northwest, and assumed it was the channel that the Belgian expedition was to head for five seasons later.

Two decades before Larsen's voyage, the German Edouard Dallmann, in the sealer *Grönland*, had sailed down the west side of what was then known as Palmer Land, named after the American Nathaniel Palmer, who had first sighted it in the sealer *Hero* in January 1820. Attaining 65° S, beyond Anvers Island, Dallmann discovered the Kaiser Wilhelm I Islands and the southern end of the island that de Gerlache would later name Wiencke Island. Seeing nothing but sea to the east between there and Booth Island, Dallmann believed he was looking at a strait leading to the Weddell Sea and named it after the German chancellor, Bismarck.

Effectively, these discoveries and Larsen's belief that the French admiral's Orleans Channel existed, expressed on an 1895 map of Larsen's voyage by L. Friederichsen, were all that was known to de Gerlache as he set off in the *Belgica* on his expedition in 1897. The Belgians' discoveries disproved the existence of the two straits, but recreated the enigma of a supposed strait further south when they sighted a 'vaste baie ou détroit' at 65° 30' S beyond the end of the Penola Strait (as the southern part of their Lemaire Channel was renamed). On the chart published after the expedition, the first officer added the question 'Détroit de Bismarck?'.

Trapped in the ice for a year aboard the *Belgica*, the Belgians escaped to reach Punta Arenas in the spring of 1899, just as the way was opening for government backing of the British and German national expeditions that would head for the other side of Antarctica.

Two other men, Dr Otto Nordenskjöld of Uppsala University in Sweden, nephew of Baron Nils Adolf Nordenskjöld, pioneer of the Northeast Passage to the Pacific in the *Vega*, and the Scottish naturalist, Dr William S. Bruce, determined to be part of the European assault on the unknown continent. In the following two years (1900–01) both men succeeded in raising funds for private expeditions, in Nordenskjöld's case with the utmost difficulty, having in the end to borrow the greater part of the cost himself.

Bruce played a crucial part in the European initiative by discovering the eastern shore of the Weddell Sea when he sailed south with the Scottish National Antarctic Expedition in 1902. The Swedish expedition, however, set off a year earlier with the aim of extending Larsen's discoveries on its western shore. Armed with de Gerlache's reports, Nordenskjöld was imbued with a desire to locate the Weddell Sea end of the Bismarck Strait that the Belgians believed they might have located.

The French Antarctic Expedition, initially conceived as an Arctic venture, and at first privately sponsored, did gain government support, something like one-third of its cost being voted by the Chamber of Deputies. However, that was only after its leader, Dr Jean-Baptiste Charcot, had met the cost of his ship, the *Français*, from his own pocket. Inspired by first reports of Nordenskjöld's discoveries, he had changed his goal and determined to head south. Just as he was about to start in 1903, news arrived that Nordenskjöld's ship had not returned, and Charcot offered to join in the search being organized by the Argentine and Swedish governments. Greeted at Buenos Aires with the news that the Argentine ship *Uruguay* had rescued Nordenskjöld's men, Charcot decided to try to extend de Gerlache's discoveries and locate the coasts sighted by the Russian Admiral Bellingshausen (in 1820) and John Biscoe (in the 1830s). Although Nordenskjöld believed he had discovered the Weddell Sea exit of de Gerlache's 'détroit', as Charcot learned when he met him in Buenos Aires, the Frenchman was destined to push the enigma yet further towards the Pole. Upon the strait's existence hung the question of whether the land was a peninsula or an archipelago.

More than four decades would be needed to resolve that question. After the Australian aviator, Sir Hubert Wilkins, had moved the supposed strait yet further south in 1929, the British Graham Land Expedition demonstrated, in 1937, that the land was the Antarctic Peninsula we know today.

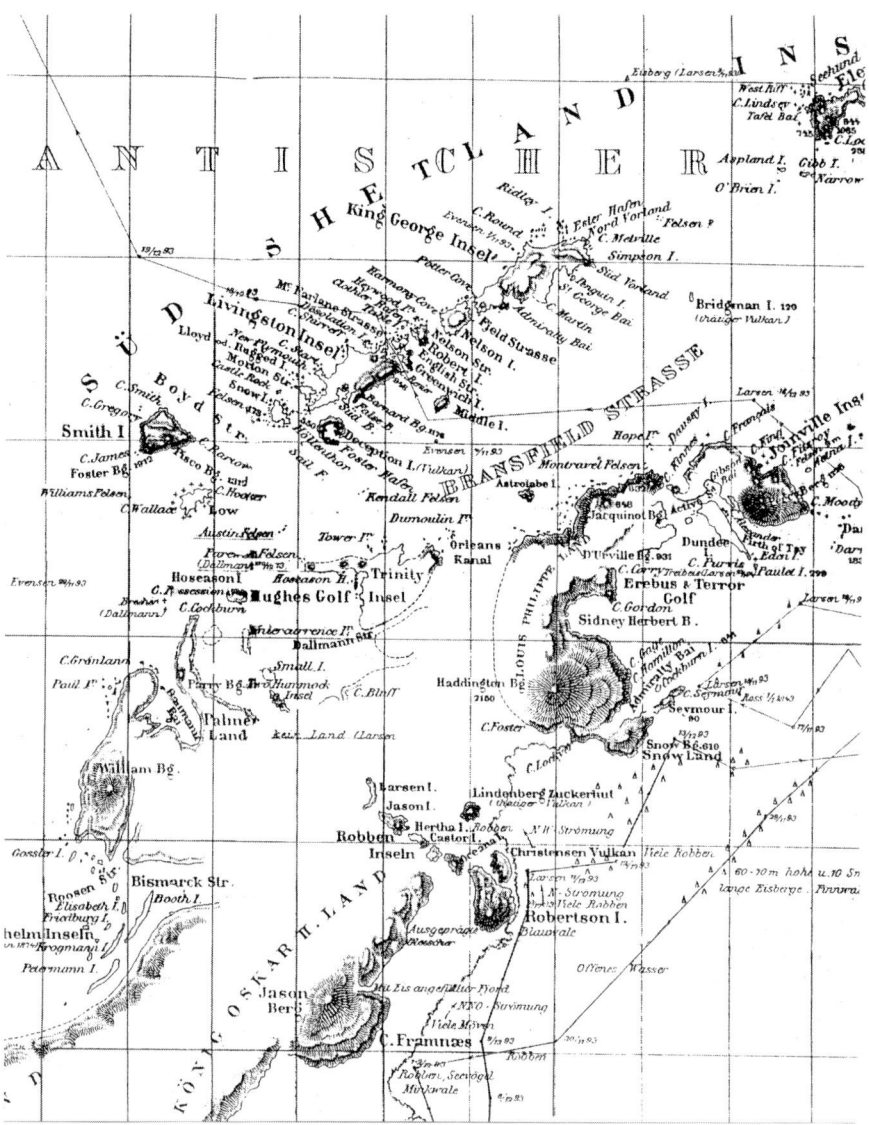

Part of L. Friederichsen's 1895 map of Carl A. Larsen's 1893 discoveries, showing the supposed Bismarck Strait and appearing to confirm the existence of Admiral Dumont d'Urville's supposed 'Orléans Channel'

Nowadays, as ice-strengthened cruise ships press ever southwards through the Gerlache Strait to Port Lockroy and on past the towering Cap Renard, then down the Lemaire Channel in the very tracks of the Belgian ship and the remarkable boat journey that five of the Frenchmen achieved and even – in years when the pack-ice permits – into the sea that sank the Swedish expedition's ship, marooning its members in three isolated locations, the story[1] of the explorers' epic journeys cannot fail to add, for the tourists on board, a dimension of extraordinary human endeavour to the awe-inspiring scenery they encounter.

Note
1 Historical notes of a more detailed nature, indicated by superscript numbers in the text, together with lists of expedition members, are placed in the Appendix. All mileages are given in geographical miles, equivalent to minutes of latitude. All three expeditions were dependent for support on carrying out extensive scientific programmes. However, it is beyond the scope of this short account to describe the methods or scientific results achieved. Those successfully accomplished, often in appallingly adverse conditions, included the first year-long meteorological observations south of the Antarctic Circle made by the Belgian expedition.

CHAPTER 1

The Belgian Antarctic Expedition 1897–1899

Moonlit photo of the *Belgica* trapped in the pack-ice in the winter of 1898

Clockwise from top left:
Lt Adrien de Gerlache de Gomery, Commandant; Dr Frederick Cook, Surgeon; Roald Amundsen, 2nd Officer; Lt Georges Lecointe, 2nd-in-Command and 1st Officer

The Belgian Antarctic Expedition 1897–1899

Following the VIth International Geographical Congress in London in 1895, the long-drawn-out 42-month campaigns needed in London and Berlin before minimum funds for a major expedition were assured had allowed two men from Europe's youngest nations, each almost single-handedly, to launch the first purely exploratory and scientific expeditions to the Antarctic since Ross's day. By surviving the long Antarctic winter night, they would lift the curtain of ignorance that hung over so many of the discussions about overwintering expeditions.

The first of them was a Belgian naval lieutenant, Adrien de Gerlache de Gomery. Smitten with Antarctic fever in the 1880s, he had applied to join the Swedish Baron Nordenskjöld's still-born Antarctic expedition in 1891, but never even had a reply. In January 1894, his ambition took a first step towards reality when he persuaded Ernest Solvay, founder of the Belgian chemical industry, to promise £1,000 (BFr25,000) for an expedition of his own.

That same month at the other end of the world the whaler *Antarctic* docked at Melbourne, bound for the Ross Sea in search of whales. Carsten Borchgrevink, a Norwegian schoolmaster working in Australia, whose wife and mother were both English, persuaded Captain Kristensen to take him on the voyage as an ordinary seaman without pay. Sailing in September of the same year, he fulfilled a dream by being one of the first party to set foot on the Antarctic continent on 23 January 1895. Or rather, the party was credited with being the first, for it had not then been learned in Europe that Captain John Davis, the American sealing skipper from New Haven, Connecticut, had sent men ashore on 7 February 1821 in the north of Graham Land, near Cape Sterneck, the eastern headland of Hughes Bay.

Commercially the *Antarctic*'s voyage was a failure, but news of the new landing reached Europe that summer and suggested one of the goals for the expedition the 29-year-old Belgian had in mind. About that time he had obtained leave to sail aboard the Norwegian sealer *Castor* (Larsen's third ship on his 1893–94 Weddell Sea voyage) in order to gain experience in the pack-ice east of Greenland. He sailed in company with Larsen's now-famous ship the *Jason*, carrying in her crew none other than the 23-year-old Roald Amundsen, likewise bent on gaining experience among the ice-floes.

For his part Borchgrevink, having failed to gain Australian support for an expedition of his own, was on his way to London. Once there he was invited to address the International Geographical Congress on the day the resolution was passed that opened the way for Markham's and Neumayer's aspirations for the British and German expeditions. Describing his experiences, Borchgrevink then spoke of the plan he had in mind:

> I made a thorough investigation of the landing-place, because I believe it to be a place where a future scientific expedition might safely stop even during the winter months. Several accessible spurs lead up from the place where we were to the top of the cape, and from there a gentle slope leads on to the great plateau of South Victoria continent . . . I myself am willing to be the leader of a party to be landed either on the pack-ice or on the mainland near Coulman Island, with ski, Canadian snow-shoes, sledges and dogs. From there it is my scheme to work towards the south magnetic pole. Should the party succeed in penetrating so far into the continent, the course should be laid, if possible, for Cape Adare, in order to join the main body of the expedition there.

Some two months after the congress, de Gerlache arrived back in Belgium from Norway. Undeterred by Borchgrevink's hopes, he began the campaign to raise support for the expedition he had dreamed of leading. Winning over the Royal Belgian Geographical Society by the end of 1895, a national appeal was launched in the new year. By June 1896 it had produced new promises of £3,000. At this point the government came forward with £4,000, more than doubling the fund, and the first practical step could be taken. Armed with that promise, for so far no actual cash was at his disposal to take to Norway, de Gerlache borrowed the £2,700 required to buy the whaler *Patria*, which he had first encountered off Greenland. Fortunately for him, the poor season forced the owner to accept far less than he might otherwise have demanded for the vessel.

Accompanied by Lt Emile Danco, released from the artillery to act as magnetician, de Gerlache remained in Norway to supervise the overhaul and modification of the ship at the Christensen wharf at Sandefjord. The required improvements included a new boiler and propeller, along with the installation of laboratories and the oceanographic gear donated by the Danish navy at the instigation of Commander Christian Wandel, their hydrographer. The work was interrupted by the winter, and despite coping with a worldwide correspondence in search of counsel and funds, de Gerlache was able to learn something of the use of skis and snow-shoes.

Guided and advised by Arctic explorers Nansen, Sverdrup, Johansen and the Austrian Weyprecht (of the Tegetthoff Expedition), the Belgian lieutenant had the ship ready by June 1897 and renamed her *Belgica*. After visits from Nansen and Sir Clements Markham, president of the Royal Geographical Society, de Gerlache sailed from Sandefjord on 26 June with Danco and Amundsen, his newest volunteer, along with two mechanics and four sailors.[1]

After picking up instruments and supplies at Frederikshavn in the north of Denmark, a rough four-day voyage brought them to Vlissingen (Flushing). There, on 2 July, they were met by Lt Georges Lecointe who had just been released, at de Gerlache's request, to be second-in-command. Lecointe found them all totally exhausted. No one had slept at all during the voyage, or eaten for 24 hours, because the trip had taken a day longer than expected. The next day they moved up the Schelde and anchored at Antwerp.

With his small country's attention focused on its Congolese ventures, public support had been as thin as it would be for the British and German expeditions that were hardly yet more than a fervent dream for their originators, Markham and von Neumayer. In the end, the *Belgica* would only sail because the government increased its contribution at the last moment to total £6,400 of the £11,900 eventually spent.[2]

Faced with fund-raising diversions to the last, de Gerlache could only assemble the remaining scientists from volunteers prepared to contribute rather than be paid. Emile-Georges Racovitza (29), a Rumanian, was accepted as zoologist and botanist; Henryk Arctowski (26), of Polish origin, would take charge of geology, oceanography and weather observations; and Louis Bernacchi (21), working at the Melbourne Observatory since 1895, would be picked up there. He would act as physicist aboard the ship for the second season while Danco was ashore with the shore party, which would winter at Cape Adare. Before they finally sailed, de Gerlache accepted, as assistant meteorologist, Polish student Antoine Dobrowolski (25) who volunteered to join without pay.

De Gerlache had to find his own engineers and crew, and he and Lecointe would sail with no powers of discipline or legally enforceable crew agreements, such as assist the smooth running of a ship under British mercantile law. In the end the heterogeneous crew de Gerlache did get together, including several sailors Lecointe described as 'undisciplined and even dangerous', was barely adequate. It was even less so, though perhaps more controllable, after the worst of them deserted during preparations at Antwerp. Such was the nature of the locally recruited men that, when Lecointe proposed exercises in manoeuvring the ship for the better safety of all before leaving Antwerp, de Gerlache had to refuse because such men usually refused any work outside the normal conduct of their duties without extra payment.

The five Norwegian sailors understood commands from Second Mate Amundsen, and de Gerlache had picked up some of their language. But the rest conversed in French. When the Belgian doctor withdrew, the problem was resolved by the fortuitous arrival of a telegram from Dr Frederick Cook (32) of Brooklyn, lately on Peary's first North Greenland Expedition, offering to contribute if he could join. De Gerlache cabled him to join them in Rio. Cook only spoke English and a smattering of German, so a third language was added which, at first, only de Gerlache and Third Officer Jules Mélaerts could understand.

All was finally ready by 16 August 1897 and, after a grand official send-off and the playing of the national anthem, the 35-metre whaler, dressed overall, sailed amid a myriad yachts and the sounding of sirens, only to have the boiler water pump break down once she was out to sea.

When the ship put into Ostend, already a month behind schedule, King Leopold II, who had hitherto held aloof from the expedition, came aboard without advance warning. Having inspected everything with minute interest, he addressed the whole company, which was some slight compensation for their troubles to date but of no avail in stemming further crew problems, which would break out twice more before they reached the Antarctic.

One of the mechanics fell sick, while the bosun and carpenter both resigned because they could not prevent sailors continually going ashore without leave. To replace them, de Gerlache only managed to get two seamen from Antwerp.[3] The working complement excluding the scientists was just 19, comprising de Gerlache, the three officers, two engineers, a steward, a cook, two mechanics and nine sailors, of whom Tollefsen, the oldest Norwegian, was appointed bosun.

Just as they were ready once more to depart, a south-west gale sprang up.

When it showed no sign of easing after three days, de Gerlache decided he must get a tug to tow the ship. So it was that, finally, on the evening of 23 August, they weighed anchor and lurched into the roughest of channel seas behind a tug, only feeling it safe to cast off when they were near the Isle of Wight. The weather remained tempestuous for 10 days, so they didn't reach Madeira until 11 September.

Sailing again on 14 September, de Gerlache and Amundsen began to sort out the supplies and equipment to be landed at Cape Adare, for the plan at that time was to use the southern summer of 1897–98 to explore southward in Weddell's tracks in the sea named after him, and then sail to South Victoria Land, disembarking de Gerlache, Amundsen, Danco and Arctowski to winter at Cape Adare and explore inland while the ship reprovisioned at Melbourne and picked them up early in 1899, after carrying out oceanographic work in the Pacific.

What with taking five more weeks to reach Rio, where they picked up the enthusiastic Dr Cook and enjoyed eight days of receptions and festivities, the *Belgica* would not reach Punta Arenas in the Magellan Straits until 1 December 1897.

At Montevideo, de Gerlache had to sack the cook and the Swedish replacement he took on (with near-fatal consequences later on) fell ill the day they sailed and had to be put ashore at Punta Arenas. The ex-foreign legionnaire steward, Michotte, volunteered to take on the job. Within three days of arriving at the Chilean port, fresh trouble broke out and a chaotic period ended with de Gerlache bringing an armed police detachment aboard to restore order. Two Belgian sailors were dismissed, along with one of the mechanics. (When eventually related in Europe, the story of the voyage out certainly did nothing to deter Markham from his conviction that the British expedition ought to be a naval one.)

About to plunge into the unknown, de Gerlache was left with a working complement of just 14 excluding Danco, the doctor, Racovitza, Arctowski and the young Dobrowolski.

The truth now bore in on the Belgian leader: they were at least five weeks late and the deck crew was dangerously small and inexperienced. Four of them were novices, 21 years old or less, with only Tollefsen and Johansen competent to lead each watch. It was only safe to sail further by virtue of the ship being equipped with Cunningham's patent furling topsails, which could be handled entirely from the deck without sending men up to the topsail yard at all.

On top of this, de Gerlache was faced with the need to accept offers of fresh meat and coal (despite picking up 100 tons shipped out from Belgium) and so had to make the diversion through the difficult Beagle Channel, which Darwin had sailed through in the 1830s, to pick them up at Lapataia. The diversion was to cost them dear.

Leaving Punta Arenas on 13 December 1897, and calling at Ushuaia, the *Belgica* reached Lapataia on 23 December. There the difficulties of embarking another 45 tons of coal, mostly stacked on deck, and some beef carcasses, delayed their departure until New Year's Day 1898. Headed for Harberton with mail and an Anglican priest, the Revd Thomas Bridges (or rather, Mr Bridges, as the Argentines all knew him), on his way to visit a fellow priest and trader, de Gerlache seized the last opportunity to replenish their fresh water. But the attempt to enter an uncharted bay in darkness nearly cost them their lives after the ship ran aground. For long and perilous hours, as a furious storm blew up, they were in danger of being completely wrecked until a large wave floated the ship off without serious damage. Unable to enter Harberton, the Belgians rode out the storm in Port Toro on the eastern shore of Navarino Island. Eventually, on the evening of 3 January, they tracked back to Harberton. After three days, and still unable to completely replenish their fresh water tanks, the *Belgica* left for Staten Island, Argentina. Arriving there on 7 January, a further week was lost, but they were able to pick up enough stored rainwater to use for drinking water until they could obtain ice for melting.

At last, on 14 January 1898, l'Expédition Antarctique Belge, so nearly finished before it really started, stood out into the Atlantic and headed the ship's bow south, past Cape Horn towards the South Shetlands and the first Antarctic lands beyond them. The opening raid upon Nature's white wilderness had got off to a shaky start.

As it was already clearly impossible for them to reach Victoria Land that season, de Gerlache decided they could only visit Hughes Bay and try to get through from there to the Weddell Sea, going on afterwards to reach Melbourne in May and stopping there for the winter. They would then land at Cape Adare the following season – it should take no longer than the original plan, but the idea of wintering on the Cape would have to be abandoned.[4]

Meanwhile, unbeknown to them, in London Sir Clements Markham had made the first announcement of Borchgrevink's expedition to the same area that de Gerlache's wintering party was aiming for. After failing to get Nansen's support in his homeland, the Norwegian had persuaded London publisher Sir

George Newnes to back his venture with almost three times the amount raised for the Belgian expedition.

As Friederichsen's 1895 map of Larsen's discoveries shows, in the north there was no obvious land connection between the discoveries of Biscoe, Dallmann and Evensen on the west side and those of Dumont d'Urville, Ross and Larsen on the east. Between them in the north there lay Trinity Land, as the Admiralty chart called it, discovered by the English sealer Edward Bransfield in 1820. To the east of Trinity Land lay Dumont d'Urville's Orléans 'channel', and to the west of it Hughes Bay, discovered by the Connecticut sealing skipper in 1821. South-east from there, Larsen had declared that there was no land, and the island mountains he showed trending north-west from Robertson Island seemed to support his theory. To emerge beside those islands was the goal de Gerlache therefore had in mind as his ship thrust south into the Drake Strait. He believed that the bay was the entrance to a channel through which he could steam or sail to the Weddell Sea.

Six weeks of the season remained to them. A week later, after crossing the Drake Strait in moderate weather, stopping only for one day to make the first-ever line of soundings to chart the profile of the sea-bed, they were running for shelter in a fierce sea in the Bransfield Strait beyond the South Shetlands.

It was then that the Antarctic claimed the first life in the new assault on her shores. The coal stacked on the deck shifted and blocked the scuppers, so the men were put to work to shift it below. Despite a warning from Amundsen, the young seaman Auguste Wiencke swung himself outside the enclosed rail to clear a scupper outlet, and in an instant was swept away by a wave. A valiant attempt by Lecointe to save him failed, and the shadow of the popular lad's death hung over the ship as they sheltered in the lea of Low Island that night, 22 January 1898.

In the manner of the quick-change Antarctic climate, the next day was perfect. Starting with the sighting of Cape Neyt (named after their first cash donor) in about the 64° S 62° W position of the possible island outline shown on the Friederichsen map, and the discovery on 24 January that the gulf was but the entrance to the strait they sought, they spent a week mapping its coasts as best they could, hampered by the absence of stars and moon, and the masking of the sun by the mountainous shores. Making the tenth of their exploratory landings, this time on Brabant Island, Danco fell into a crevasse and, after the rope broke, was only saved by his ski getting stuck across a narrower part of it. During the eight days they were ashore, Lecointe took the *Belgica* over 60 miles

10 QUEST FOR A PHANTOM STRAIT

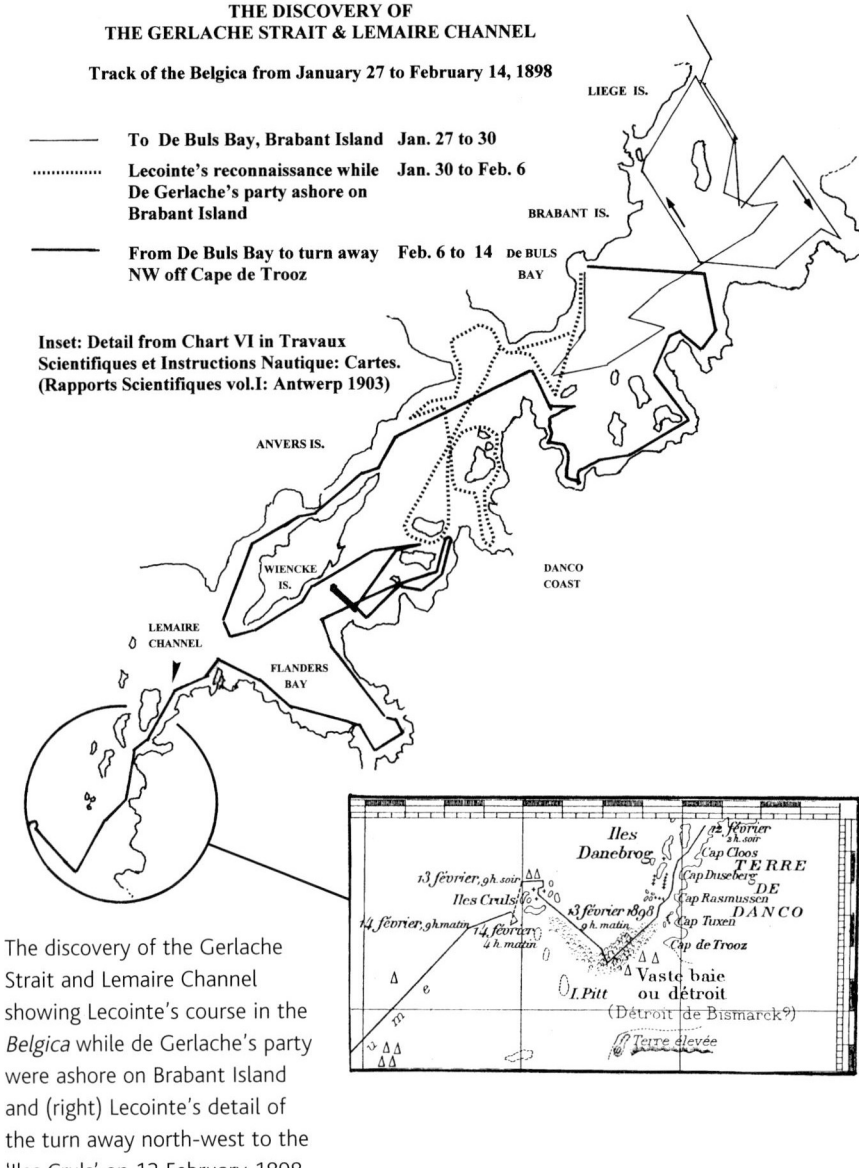

The discovery of the Gerlache Strait and Lemaire Channel showing Lecointe's course in the *Belgica* while de Gerlache's party were ashore on Brabant Island and (right) Lecointe's detail of the turn away north-west to the 'Iles Cruls' on 13 February 1898

further south into the channel, finding that it continued south-west with no sign of turning towards the Weddell Sea.

Out at the far end after four days, on 11 February, the Belgians established that Dallmann's Bismarck Strait was a large bay, which they named Baie de Flandres. Following the coast a further 25 miles southward, they passed inside the Kaiser Wilhelm I Islands (respectively Booth, Krogmann and Petermann Islands), also discovered by Dallmann in 1874, without recognizing them. Imagining that he had discovered them, de Gerlache renamed them Wandel, Hovgaard and Lund Islands, after his Danish supporters, calling the group the Iles Dannebrog. Beyond the last of them at 65° 15′ S they encountered impenetrable ice. It was 13 February 1898. This spelled the end of their geographical achievement. In 22 days they had redrawn the map and, along a 90-mile stretch, begun tracing the western coast of the Antarctic Peninsula which runs a further 500 miles to the south. The unfinished charting of that stubborn coast was to inspire expeditions for decades to come in the new century, not least because, as Lecointe's map shows, before turning away to the north-west they had seen what he described as a 'vaste baie ou détroit (Détroit de Bismarck?)'. It was to leave an enigma that permeated the thinking of the leaders of the next two expeditions.

For the Belgians a frustrating two weeks followed, tracking the ice ever to the south-west looking for a way through. Sighting Alexander Land a tantalizing 40 miles away, its mountains rearing up over the horizon, they could see no way of getting near it. For the wintering party it was a very poor alternative to Cape Adare.

On the last day of February, when ships usually sought warmer waters, they had reached $84\frac{1}{2}°$ W. On being told that there was suddenly a clear way to the south, de Gerlache quickly came up on to the bridge. The wind was blowing from the east-north-east so they could sail south or north with equal ease. Already, three days before, the scientists had spoken out against attempting to push further south this late in the season, but after a short conversation in which both men recognized the risks, the possibility that, like Ross, they might break through into an open sea clearly outweighed the dangers for Lecointe as much as for de Gerlache. 'With profound excitement [joie]', Lecointe later wrote, 'I gave the order to head the ship southward.'

In that rapid decision to seize the hoped-for opportunity despite the lateness of the season, the two men seem to have forgotten that Ross had taken his similar decision almost eight weeks earlier in the season, on 5 January 1841.

The wardroom of the *Belgica*: (left to right) Arctowski, Lecointe, Racovitza, de Gerlache

After a day ploughing through relatively fragmented ice, Ross, seeing a watery sky to the south, had led his ships into thicker ice, from which they emerged three days later. No such sky greeted the advance of the *Belgica*. Either forgotten, or discounted too, was the fact that late in February Ross had found the pack-ice hard against the land and could find no way southward into it.

But even if the two Belgians remembered that, as they took their risky decision, there was another factor in their minds on the bridge that day. Of the expedition funds only £640 remained (surely firm evidence that their budget always was grossly inadequate for what they had set out to do), and even if the staff contributed the whole amount of letters of credit they carried, the total then available would scarcely be enough to put the ship in good order and reprovision her. It was certainly not enough to pay the extra hands needed to bring the crew up to strength.

Rather pessimistically they thought that the achievements so far would not win them further government support. So, in their eyes, if they returned to

South America, the expedition would be over. If they pushed south and broke through, there was at least a chance that some tremendous discovery on a par with Ross's might justify everything after all. They might just, as Murray on the Discovery Expedition would surmise later that year, emerge into the Ross Sea in time to land the wintering party at Cape Adare.

Three days later they were stuck fast at 71° 20' S 85° 30' W, further south than the point reached by James Cook in 1774, but some 400 miles further east. The rest of their story would be the achievement of the scientists: the first year-long series of scientific observations south of the Antarctic Circle, faithfully pursued through two epidemics in which the will to survive came near to extinction in a world of ice stretching to every horizon. From late May the light was dim at midday, alleviated only by the glow of the northern horizon and varied occasionally by a day of bright moonlight, until the growing return of daylight in mid-July.

The expedition's long imprisonment was to show, however, that the darkness and the weather were not their main enemies. Nor would the pack-ice be the menace many feared. Only once, the morning after Midwinter's Day, did its pressure make them prepare to abandon ship.

As in the Arctic, whenever men had been denied fresh food for more than about two months, the dreaded scurvy made its appearance, but in this case complicated by acute anaemia and circulation ailments in many of the party. The little band of men on the beleaguered ship faced that ordeal twice before their escape. It first struck after two and a half months locked in the ice, but initially the symptoms were not the usual ones. De Gerlache developed constant headaches, and two weeks later Danco suffered from serious heart trouble. His condition rapidly worsened and by 5 June he was dead.

Danco's death didn't really suggest diet deficiency because de Gerlache had taken him on knowing he had a history of chest weakness. But Cook was already worried, on 20 May, about the general decline in health and by early July Lecointe had developed the classic symptoms of scurvy: swollen gums and stiff joints. A few days later Cook insisted that Lecointe, who was now sinking fast, should force himself to eat raw penguin steaks. It was almost too late.

Three months earlier, in late March, the fresh beef appears to have run out. In the meantime they had lived largely on a Norwegian version of tinned pemmican recommended by Nansen – 'made of minced cats' Amundsen told them, according to Lecointe. If Amundsen was not joking, it hardly sounds like the pemmican later taken on his own South Pole expedition, which was called

Cook's tent photographed on the abortive sledge journey on 30 July 1898. The asymetrical shape was vulnerable to strong wind on the entry side

kjøttboller and *kjøttpølser* (meatballs and sausages). The Belgians also had some tinned Australian rabbit, which does sound like the 'Australian pemmican' Nansen had used. De Gerlache had found it all quite palatable when freshly prepared in Norway. Now it all tasted repulsive, the tinned rabbit being worst of all. The experience of tinned food going bad aboard the *Discovery* four years later suggests that this is what may have happened here.

This monotonous diet was endured despite the fact that, on 26 March, they had set about hunting seals and penguins in an energetic campaign that Lecointe described as 'days of carnage'. Five weeks later, faced with the crew's growing discontent over the diet, Lecointe confronted de Gerlache, pointing out that he had never been told how much fresh meat they had amassed and demanding to know why it was not on the menu. To his amazement, the commandant's explosive response was that of course there was plenty, but what would the press say later when it got out that they had eaten seals! After a long argument, de Gerlache produced a list of varied menus the next day. The trouble was that Michotte had been unable to make the raw meat palatable, and no one could bring himself to eat the seal or penguin dishes he prepared. Of the two, the penguin was evidently the less nauseating.

Just as Lecointe, who, at his worst, had been unconscious for hours at a time, made his dramatic recovery in the second half of July, Michotte at last found out how to remove the oily taste from the seal meat. Yet, even then, de Gerlache, Mélaerts and two others refused to eat it. It was too late to prevent Johansen joining the ranks of stiff-limbed sufferers, while, disturbingly, another sailor became hysterical. Like Lecointe, Amundsen and Cook recovered well, and joined the others in hunting seal daily, which Lecointe implies most of them ate in large quantities (though Cook wrote that Lecointe only ate penguin from then on).[5]

Nevertheless, by mid-September, all the deck crew, except the teenager Koren, became afflicted with acute anaemia, and de Gerlache started having headaches again. Despite fresh raw meat being almost their staple diet that month, and the table being 'liberally supplied with fresh steaks', de Gerlache, Mélaerts and Michotte all developed the usual scurvy symptoms. Finally Lecointe, who claimed that he did in fact continue to eat seal, developed the symptoms in November, becoming too weak to carry on the magnetic observations. Eating as much extra seal and penguin as he could manage, he had recovered by the end of the month. De Gerlache, on the other hand, continued to deteriorate and, on 4 December, he and Lecointe drew up a document appointing Mélaerts and Somers to succeed them if they should die, it having been agreed with the Royal Belgian Geographical Society, their principal sponsors, that command should remain in Belgian hands.

With some justification, Lecointe must have feared that the scourge would return despite a diet of fresh meat. Happily it did not, for, by mid-January 1899, all but two men were fit enough to join in operations to cut the ship out, including the unhappy bosun. He had developed persecution mania at the height of the first outbreak, and became permanently deranged in December on return from a ski-trek to an iceberg nine miles from the ship – the longest journey made from the ship.

Not surprisingly, during their long ordeal, little beyond valiant scientific observations had been accomplished. Back in May, before the first symptoms appeared, Cook had tried out his idea for a sail-driven sledge, but its centre of gravity was too high and it constantly overturned, even with him riding on it while Lecointe steered. Cook tried in vain to modify it. Then came de

Gerlache's headaches and Danco's death, and on the last day of June, little knowing that Lecointe would be at death's door and he himself seriously ill within three weeks, de Gerlache proposed that Lecointe lead a 15-day sledge journey to the south when daylight returned, taking Amundsen and Cook with him. Arctowski had obtained soundings of 420 metres through the ice where they were, and the commandant believed (correctly, as today's maps show) that this meant land was not far to the south.

The trio's health being restored, they set off on 30 July 1898 to make the first-ever sledge journey on Antarctic ice, using a small tent designed by Cook. The next day they were fog-bound and built themselves an igloo in which they spent the second night. The third day they advanced by moonlight, stopping every 20 metres to take a compass reading. After spending the night in the tent with the ice breaking around them, a mirage made them think the ship was in open water and they started back, fearing de Gerlache couldn't see them; but he had, and he later sent out sailors to meet them. They abandoned all their gear and hurried in, deciding that de Gerlache's plan was far too risky.

Lecointe had found the experimental tent too cramped by half. Cook later modified it, but the Belgian, while finding no major fault, still thought it too small. Curiously, Amundsen, a taller man than Lecointe, made no complaint about it. It only weighed 12 lb, but it was anything but streamlined if the wind shifted to the entry side once it was pitched. They never slept in it again, for the ice remained absolutely solid around the ship until Christmas, by which time all possibility of reaching the Ross Sea to use it that season had disappeared.

By the middle of January 1899, after being thwarted in an attempt to blast a way out to one piece of open water, they located a long belt of ice one metre thick leading, in little over a mile, to another open pool where they began to cut triangular pieces out of it from the water's edge inwards towards the ship. After four weeks the ship was free and they made 20 miles' progress northwards only to have the ice close in around them once more. As the men aboard the *Belgica* woke the next day, Borchgrevink's party aboard the *Southern Cross* first sighted Cape Adare after a long struggle to get through the pack-ice.

By the time Borchgrevink and his shore party were installed in their hut early in March, the *Belgica* was being carried west-south-west again, and the survivors were wondering once more if they would after all emerge in the Ross Sea, as de Gerlache and Lecointe had believed they might when they took the fateful decision to push into the ice a year before.

Fortunately for them it was not to be, and after six more days they were

THE BELGIAN ANTARCTIC EXPEDITION 1897–1899

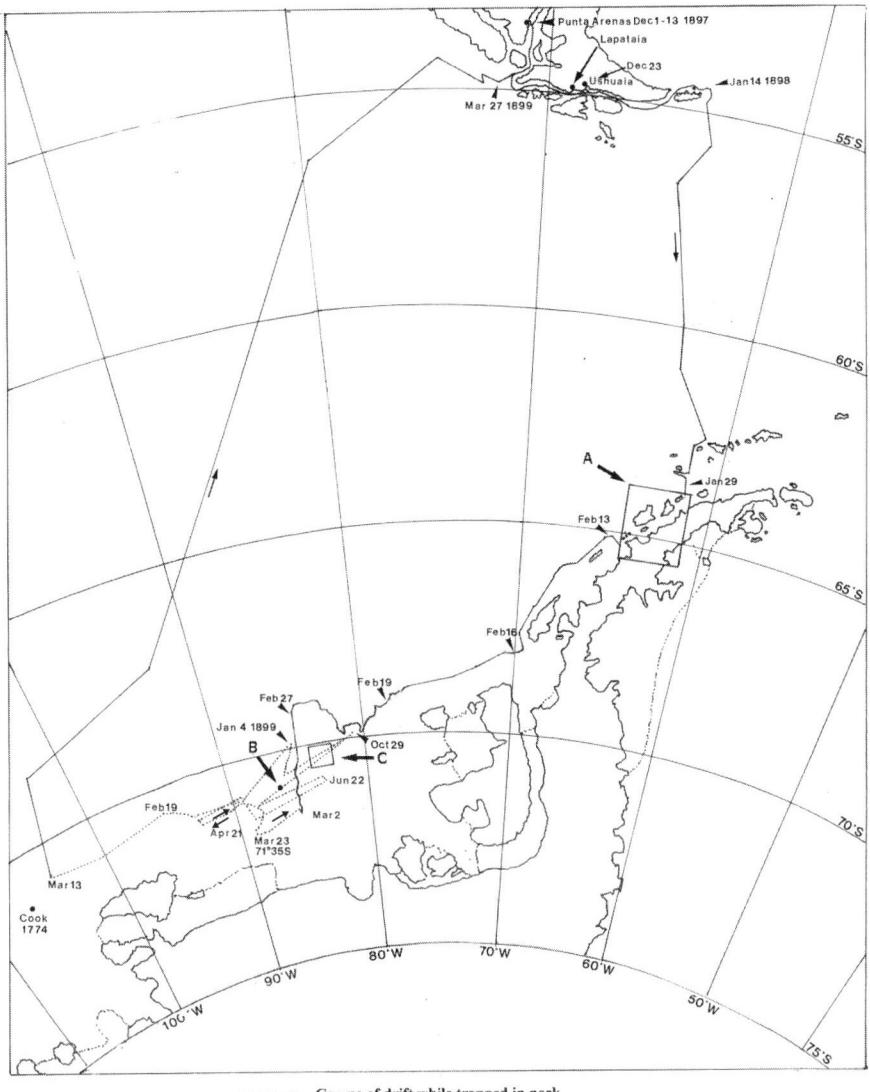

.... = Course of drift while trapped in pack shown much simplified from Jun 26, 1898 to Jan 4, 1899)

A = See map of Gerlache Strait and Lemaire Channel.
B = Start of abortive sledge journey on Jul.30, 1898
C = Vessel was in this area from Aug.20 to Oct.10 and from Nov.10 to Dec.18, 1898.

The track of the *Belgica* from 13 December 1897 to 28 March 1899; from charts by Lt Georges Lecointe published with the scientific reports free. Two weeks later, on 28 March 1899, they dropped anchor in Punta Arenas, evoking much consternation for they had long since been given up for lost.

The ship would have to be overhauled before she sailed for home, and there was not enough money left for that. So the two scientists and their assistant were sent back by steamer with part of the collections to start work on classifying them and the physical science records. Cook was released and, after further study of the aboriginal indians, returned to the United States. The unfortunate bosun, once ashore, could not be persuaded to reboard the ship. Amundsen, who had handed de Gerlache a formal written resignation when he learned that Mélaerts, and not he, would be promoted if either de Gerlache or Lecointe died, escorted him back to Norway.[6]

The remaining stock of the detested tinned food was sold and enabled the repair bills to be paid. De Gerlache then took the ship round to Santa Cruz, put Lecointe ashore with a party to make magnetic observations in the mountains, and return within a month. Johansen and Koren were in the party, but had to return after Johansen was taken gravely ill; he had to be left at Santa Cruz. Lecointe himself got back too late and, as arranged, then took himself back to Belgium, while de Gerlache and Mélaerts, who could not afford to buy coal, left Buenos Aires on 4 August with the engineers, Michotte and just four deck hands[7] to sail laboriously across the Atlantic. Forced up to the Newfoundland banks by contrary winds, it took them two and a half months to reach Boulogne, where they were joined by Lecointe and the scientists. Sailing into Antwerp on 7 November 1899, they received an ecstatic welcome, first there and then in Brussels.

CHAPTER 2

The Swedish Antarctic Expedition 1901–1903

The shore party on board the *Antarctic*: (left to right) Bodman, Akerlündh, Nordenskjöld, Jonassen, Sobral, Ekelöf

The southern party ready to start: (left to right) Nordenskjöld, Sobral, Jonassen. Note the absence of skis and the Falklands sheepdog behind the four Greenland dogs

The Swedish Antarctic Expedition 1901–1903

Otto Nordenskjöld's Swedish Antarctic Expedition left Gothenburg on 16 October 1901. For Nordenskjöld, nephew of the first man to navigate the Northeast Passage, the departure was the climax of a huge personal effort, dogged from first to last by lack of financial support, which he only surmounted by loans he would spend the rest of his life repaying. His ship was that same Norwegian sealer, the *Antarctic*, in which Borchgrevink had first seen the southern continent. With a largely Norwegian crew in the hands of the by then renowned Captain Larsen, he was bound for the very route that Larsen had pioneered on the west side of the Weddell Sea.

At Falmouth, where they were picking up coal, 14 Greenland dogs were brought aboard by the Norwegian Ole Jonassen who had sledged with the Duke of Abruzzi on his North Pole attempt two years before. Weighing anchor on 27 October, they had crossed the Equator by 24 November and arrived at Buenos Aires on 14 December, having lost 10 of the dogs in fights or from tropical heat.

Having previously agreed to take an Argentine naval officer with the ship in exchange for the co-operation of the Ano Nuevo Observatory, off the north shore of Staten Island, Argentina, Nordenskjöld soon found himself faced with a demand that the appointed officer, a Lt José Sobral, be included in the shore party. Despite having designed the base hut for a six-man party, an interview facilitated by a common knowledge of German persuaded him to agree. It was a fortunate decision, for the Argentines promptly offered unlimited help, a promise that was to be redeemed in a manner none of them could have foreseen.

Joined by the American artist Frederick (Frank) Wilbert Stokes, who had been with Peary on his first two Greenland expeditions and made a handsome contribution to Nordenskjöld's funds, the *Antarctic* slipped her moorings on 21 December and headed for the Falkland Islands to buy sheepdogs. Only managing to obtain four, the expedition finally sailed southwards on New Year's Day 1902.

After calling at the Ano Nuevo Observatory, and finding that the key instrument for co-ordinating their own magnetometer with an established magnetic base had not even been commissioned, they headed south past the fearsome Cape Horn on 5 January. Their aim was to pass through the 'Orléans Kanal', which Larsen still believed to run through to the Weddell Sea between Dumont d'Urville's Louis Philippe Land and the Belgians' Danco Land further west. The goal then was to establish a winter station as far south as possible on the western coast of the Weddell Sea, preferably beyond Larsen's farthest 1893 sighting at 68° 10' S.

Safely across the Bransfield Strait, they had sailed west into the Orleans Channel to find that after 75 miles it emerged not into the Weddell Sea, but into Hughes Bay at the entrance to Gerlache Strait. They had made their first discovery – the two lands were one and the same. To reach the Weddell Sea they would have to try the sound separating Louis Philippe Land from Joinville Island, which had been blocked by ice when the French Admiral Dumont d'Urville discovered it in 1838. Never since navigated, it proved to be entirely free of ice, and they passed through on 15 January to emerge in Ross's Erebus and Terror Gulf, where Nordenskjöld disembarked on the eastern shore of Paulet Island in search of fossils. On the way through the sound, as though guided to the very places that would figure in future crises, he had pointed out a sheltered beach at the foot of a magnificent glacier in Hope Bay (as they later named it) on the west side of the sound as very suitable for a depot.

They set up their first emergency depot on Seymour Island, 40 miles south across the gulf, and groping their way southward in fog, the Weddell Sea pack-ice brought them to a halt at 66° S, 20 miles east of Larsen's Cape Framnaes. When the weather cleared they could see the towering ice shelf either side of the cape – as high as Ross's Barrier on the other side of the world – and above it glimpses of mountain peaks beyond the nearest, which was Larsen's Mt Jason. Believing the cape might be suitable for a depot, Nordenskjöld tried skiing to it, but after two hours abandoned the attempt, for it was clearly beyond them to sledge all the coal, huts and supplies over such a distance.

Determining to try again further east, the retreat northward to ice-free waters off Ross's Cape Lockyer led to their first real discovery. After a vain attempt to approach Robertson Island, Nordenskjöld had Larsen lay the ship alongside the fast ice further north to see if it was suitable for sledging. From there they sighted distant peaks, exactly where Larsen had believed the Orleans Channel to emerge from the north-west, and Nordenskjöld realized he had discovered the east coast of de Gerlache's Danco Land.

THE SWEDISH ANTARCTIC EXPEDITION 1901–1903   23

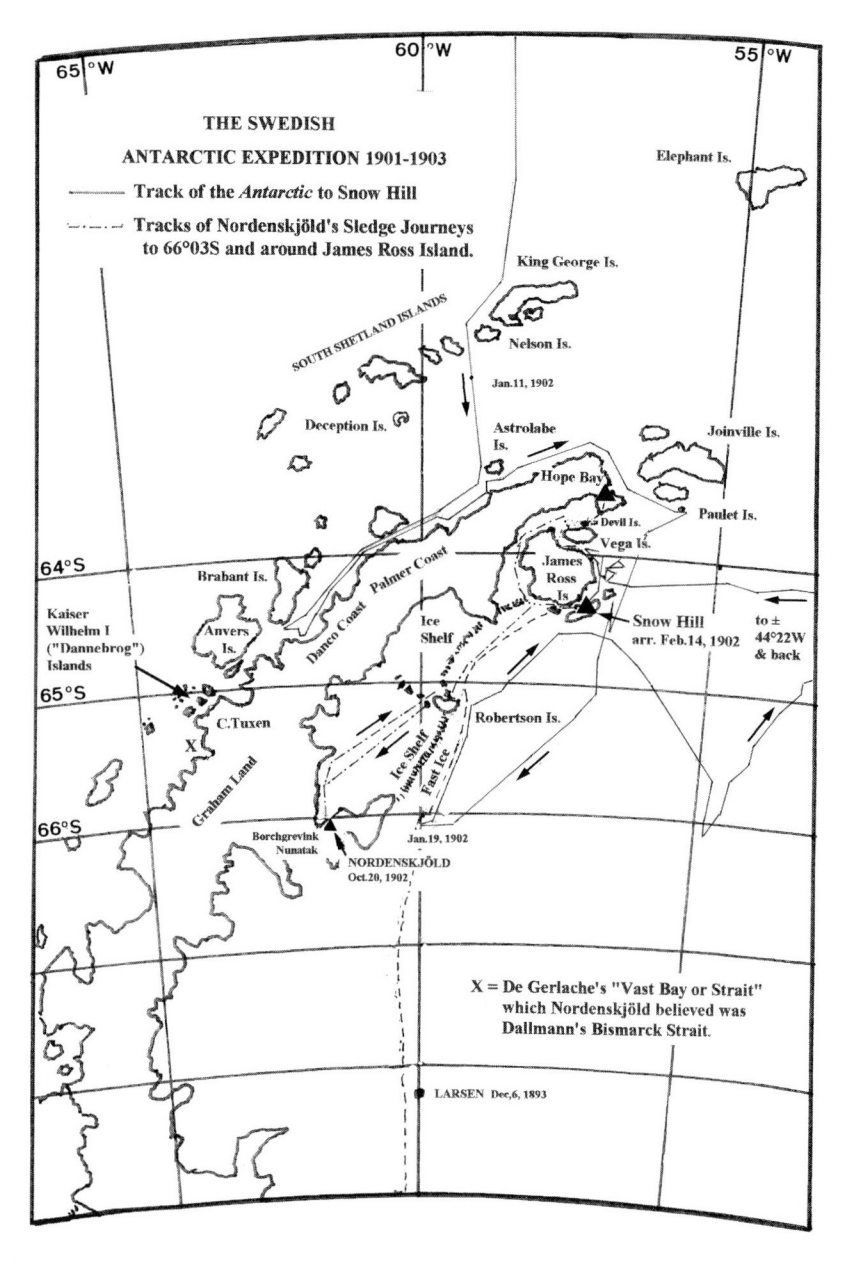

The Swedish Antarctic Expedition 1901–1903: tracks of the *Antarctic* to Snow Hill and Nordenskjöld's sledge journeys

He was not allowed long to contemplate the discovery, for a storm swept the ship away eastward that evening. Four more times Larsen tried to break through the pack-ice, once even getting beyond the 66th parallel to the south of the South Orkneys, until their dwindling coal supplies forced a retreat on 2 February 1902. The west coast of Seymour Island seemed to be the only practical choice for their base, even if it was only 10 miles south of the 64th parallel.

It took three days to extricate the ship from the pack-ice and four more for the 320-mile slog back to the island against contrary winds. Driven out of the entrance to Admiralty Sound by a blizzard, a clearer view to the west on 10 February suggested that Ross's Sidney Herbert Sound might be a safer place for their base. They were quickly disabused that evening when the ship ran aground on a sandbank; fortunately, the next tide floated them off.

Unperturbed, Stokes sat on deck to paint the glowing intensity of the clouds after sunset. Although it was just the sort of unique spectacle the American had come south to find, he had already decided not to stay, for so northerly a base would surely deny him his first artistic passion – the brilliant auroral displays he hoped would excel those in the north (a vain hope in that region, so far from the south magnetic pole).

That neatly resolved the dilemma Nordenskjöld had faced once Sobral was included in the shore party, which had meant seven men using the tiny 13 ft by 7 ft living room in the hut designed for six – a decidedly bleak prospect for their comfort during the long winter evenings. The shore party would now comprise Nordenskjöld, magnetician Gusta Bodman, also handling meteorological work, doctor and bacteriologist Erik Ekelöf, José Sobral acting as assistant to Bodman, dog driver Ole Jonassen, and the 19-year-old able seaman Gustaf Akerlündh.

Back in Admiralty Sound on 12 February, fortune led them to the perfect site sheltered in an inlet on its east coast. They had arrived none too soon, for after four days, with only the main frame erected, ice arriving from the north forced Larsen to leave for a final attempt to land the emergency depot on Robertson Island. Nordenskjöld considered that vital to a second attempt to get further south the following season.

When Larsen returned on 21 February to report another repulse by the ice, the hut was barely completed. Landing another 30 tonnes of coal, the six men of the shore party were left on their own that evening as Larsen headed north with instructions to return at the earliest feasible date for another attempt to land them at or beyond Cape Framnaes.

After the calmest of passages northward across the Erebus and Terror Gulf, and back through the sound later named after his ship, Larsen's fortunes changed abruptly as a violent easterly storm broke on their passage past the coast of King George Island. Saved from shipwreck on a lee shore by Larsen's consummate seamanship, the battered *Antarctic* had struggled into Ushuaia in the Beagle Channel on 12 March with her bunkers almost empty. From there, replenished with free Argentine coal, and with the worst damage patched up, she returned to Port Stanley on 27 March for the extensive repairs needed before she could undertake her winter voyage to South Georgia.

Down at Snow Hill, as he now named the shore base, the storm that so nearly wrecked the ship had dealt Nordenskjöld another blow. Amid plunging temperatures the Greenland bitch had broken out of her kennel and left her litter to freeze to death. It was the end of his hopes of increasing his derisory dog force, so essential to any attempt to sledge south from Cape Framnaes.

Equally vital had been the depot on Robertson Island, which the ice might yet again prevent them establishing. So a plan to establish it for themselves was born and, before Larsen had reached the Falklands, Nordenskjöld had moved 500 lb of dogfood for it to a temporary depot at Tortoise Hill (its modern name) on the coast south of Ross's Mt Haddington.[1]

Taking the dogs and a sledge, they used the boat to reach the end of the sound, there to find that it ran through to the sea, separating their base from the mainland. Ross's Cape Lockyer proved to be on what is now Lockyer Island, separated from Tortoise Hill by a channel. Along the coast west of the depot lay just the fast sea ice they would need to reach the route south from Cape Foster to Robertson Island, some 65 miles from the depot, itself 14 miles from their base.

Carrying on with the scientific observations through the winter, in temperatures that fell to an average -25 °C in June, the month of their worst storm with 60-knot winds, Nordenskjöld abandoned the idea of going south in the ship in favour of making the entire journey from their base, with the ship picking them up if Larsen could get through. The dogfood would have to be advanced to Robertson Island at the start of the main journey.

However, during the winter, two of the Falklands sheepdogs had been killed in fights – an inevitable result of Jonassen's failure to tether them. So, with another dog injured and the two bitches nursing litters, they had to start with just three. The plan to move the depot all the way to the island was reduced to establishing it on a convenient iceberg on the way.

This scheme was promptly ruined at their first camp when, unbelievably, after once more failing to tether the dogs, they woke to find that two Greenland dogs had deserted, their tracks showing that they preferred the comforts of the base. Moving some supplies to the depot below Tortoise Hill, they manhauled the rest back, determining to start again with the four Greenland dogs and Kurre, the surviving sheepdog. Soon after their return all the ice disappeared from the sound, which was only safely refrozen by 29 September 1902.

By then, Nordenskjöld had reverted to believing that the ship could get them beyond Larsen's furthest south and so decided they must be back by mid-November. Since that left only 45 days for the entire journey, something had to be done to give them more time. So orders were left for the ship to pick them up at Robertson Island late in November, which he reckoned would gain them another 15 days.

The limitations for inland exploration undertaken by a small shore party were promptly displayed in the compromise now imposed on Nordenskjöld as he worked out the detail of the new plan.

There was room for only 21 days' supply of dog pemmican. That the journey could start at all depended on the fact that they would be following the seaward edge of the coastal fast ice and they would find seals and penguins to make up the deficit first at the island and then, as they headed towards the coast, at Larsen's Seal Islands, west of Robertson Island. With a load of 700 lb to haul, although that would eventually reduce, any idea of dogs being sacrificed to feed others was as out of the question as a support party to haul depot supplies. And the party could not even include a doctor.

As it was, the Greenland dogs, undoubtedly the strongest of all in the Antarctic in 1902, and fed on penguin meat all winter, saved the day. Starting on 1 October, they picked up the Tortoise Hill depot next morning. Nordenskjöld and Sobral, pulling 200 lb on the smaller sledge, could not keep up with the dogs and so had to transfer 50 lb to the dog sledge. Even then Jonassen had to ride on it to slow it to their pace.

The weight each dog was pulling amounted to almost 150 lb, and must have been more than that for the four Greenland dogs, since the Falklands sheepdog was unlikely to have exerted the same pull. That the 65-mile journey from depot to island was accomplished in five days on the move can hardly have been due to anything but a favourable surface, even allowing for Jonassen's experience, the limitations of which were to be demonstrated later.

Losing just one day in a blizzard, they arrived at the island on 7 October, there to be greeted with an unpleasant surprise. The sight that greeted Nordenskjöld on his climb to the summit of Mt Christensen once more caused him to change his plan. Seals there were in plenty, but the 100 ft ice wall they had encountered running away north-west from there was really the edge of an ice shelf that engulfed the so-called Seal Islands and stretched away in a limitless expanse to the south. There had never been any seals near the islands, nor would there be on the coast they planned to follow to the south-west. Somehow, neither Larsen in 1893, nor they the previous autumn, had seen anything of it on their passage past Robertson Island.

This meant that they would have to take all the food with them instead of leaving some at the island. Counting on finding some at Cape Framnaes – a risky assumption considering that he had not seen its coast – Nordenskjöld left a note telling Larsen he would leave news at that cape, and might return along the coast all the way to Cape Foster.

The immediate problem of getting up on to the 300 ft shelf was solved when Nordenskjöld found a snow ramp after the evening meal. It took the whole of 9 October to climb it and reach their camp on the Oceana Island nunatak.

Narrowly avoiding disaster the next day, when Jonassen capsized the sledge after letting the dogs plunge down into the huge bergschrund surrounding the Seal Island nunataks, they reached the northern side of Castor Island, the southernmost one, the following night. The discovery of a gentle slope back up to the shelf surface and a good following wind helped them make good progress on 12 October, only to be held up by a day-long blizzard the following day. Starting south-west in better conditions on 14 October, three remarkable days in the annals of sledging ensued, during which they covered some 57 miles – 19 miles a day by Nordenskjöld's estimate using his pedometer, for they had no sledgemeter.[2] For position, they relied entirely on Sobral's fixes with an artificial horizon, which the weather prevented him from using until several days later. Considering they could only move at the speed of the two men pulling on foot, not on skis, and always ahead of the dog sledge, it was a marathon effort.

Weatherwise, it was their only lucky spell south of the island. Pinned down by a blizzard for the next two days, Nordenskjöld realized that his only hope lay in a lightweight push for the peaks which, as the storm subsided, he began to see to the south above the whirling drift. All that the remaining rations would afford was an eight-day round trip. They would have to take the smaller sledge with just the bare essentials, leaving the rest in a depot.

By lunchtime on 18 October, conditions had improved enough for them to start. With the coast to the west largely obscured by drift, they were crossing the badly crevassed snout of a glacier within the hour. With all of them on foot, the three men somehow got the sledge across it, Nordenskjöld in the lead and sinking through a snow bridge up to his armpits every now and then, while the other two helped the dogs drag the load.

They crossed a wide valley in the ice, scaled another ice face by a snow ramp and dragged the sledge up a long slope of hard blue ice to reach the foot of a volcanic nunatak projecting almost 1,000 ft above the ice. After six hours of relentless going, at one point being saved by the sledge from destruction in a crevasse, Nordenskjöld stepped on to solid ground, the first man to set foot on the coast that had defied every other attempt to approach it.

Forced to move camp twice that night as a veritable tempest descended on them, Jonassen badly injured his arm and, to cap everything, the dogs broke into their pemmican bag and devoured its entire contents, putting paid to any further advance.

When the storm died down on the evening of 20 October, Nordenskjöld climbed the nunatak, which he named after Borchgrevink, to find everything blotted out by mist. The summit he stood on was just three miles south of the 66th parallel, and they had travelled 177 miles.

The next morning, leaving the others to pack the sledge, Nordenskjöld climbed it again to scan what looked like an ice-filled sound running west, flanked by mountains that dwindled into the distance where it seemed to curve northwards. Standing there with his camera, some 13 miles south of their camp of 17 October, he could see that the coast beyond him ran away to the southwest, but the view to the west held the promise of an infinitely more dramatic discovery.

Like de Gerlache, he was convinced of the existence of Dallmann's Bismarck Strait. Assuming the sound ran on to the north-west, and knowing that was the very direction of the 'vaste baie ou détroit' on Lecointe's charts which de Gerlache believed might be the real strait, Nordenskjöld concluded that the two were connected, although he was sure the passage would never be navigable.[3]

As it would so often do, the Antarctic had deceived the eye of its beholder, not a little aided by the allure of discovery. It was to fall to the French, 14 months later, to spell an end to Nordenskjöld's hopes and move the elusive 'strait' further into the southern mists, there to remain an enigma that, a quarter of a century later, would draw the Australian aviator Sir Hubert Wilkins southward to pursue its quest.

As he returned to his camp that day, 20 November 1902, Nordenskjöld faced the formidable challenge of how to accomplish the rest of his goals without any pemmican for the dogs. Starting on their precarious run back to the depot camp with the coast always invisible, Nordenskjöld was haunted by fear of missing the depot that carried their only means of survival.

However, all went well and, once arrived, Nordenskjöld recalculated the rations to see what they would be able to do. The dogs would have to be fed on the men's pemmican. An acceptable reduction in their own ration could provide the dogs with 8 oz daily. That ought to allow them to follow the coast, provided the weather held. As though in response, the next day dawned brilliantly fine, their first good day since leaving the Robertson Island camp.

By then north of the big glacier they had crossed, it was too late to see the 6,500 ft skyline at its head, and so they missed the chance of realizing that the Richthofen Valley, which Nordenskjöld mistook for the Bismarck Strait, was only a stagnant overflow from a breach in its southern wall.

Now there was the prospect of getting on to the land for rock specimens at a prominent cape 16 miles ahead of them. However, they found their hopes frustrated by a deep canyon between the ice shelf and the coast, and were forced to bear away eastward. To this day the cape bears the name they gave it – Cape Disappointment.

When 23 October dawned with more thick fog, they knew that the opportunity of following the coast had gone. Their survival depended on heading directly for Robertson Island and when, a day's run from it, a two-day blizzard ate into their reserves, they began to think they would have to kill the sheepdog to feed the others. But the next day, 30 October, was fine enough to pose no problems for their descent to the sea ice at their old camp, where plentiful seals ended their worries.

Aided by a makeshift sail and a stiff following wind, they were back at Lockyer Island after three more days. On their arrival, the weather looked menacing, so they all voted to continue despite having covered 23 miles that day. Carrying on up the sound, they arrived in a completely exhausted state at 1.30 a.m. on 4 November and roused their surprised comrades, fast asleep in the hut. The Falklands dog collapsed, doubtless permanently weakened, for he was soon killed by his erstwhile harness mates. Sobral fainted after they had eaten, and Nordenskjöld nearly did likewise while helping him into his bunk.

The Swedish university lecturer was well pleased with what they had achieved, in spite of the twin blow to his plan, dealt by the lack of seals and Jonassen's injury that had undoubtedly brought about the demise of the bag. Travelling on the sea ice some 30 miles from the land they had discovered some 65 miles of coast never seen before and gone on to locate, though not survey, 80 miles of the coast south of Robertson Island, the peaks of which Larsen had sighted nine years before. Gradually approaching the first 45 miles of it they had gone on to sledge very close to the rest, even if they could not see the last crucial stretch. Finally, they had actually set foot on it, and Nordenskjöld had seen what he believed was the elusive Bismarck Strait.

At first confident that Larsen would soon arrive, the weeks that followed turned Nordenskjöld's satisfaction into a growing suspense, punctuated for him only by a double fossil find on Seymour Island, when, on 4 December, he visited it to look out for the ship. There he had chanced upon fossilized bones which he immediately recognized as important – they proved to be those of penguins some 65 million years old – and then came upon the much-sought-after plant fossils that, until then, had eluded his greatest hope as a geologist well versed in palaeontology. Little knowing that his find was shortly to be eclipsed by his own second-in-command, the geologist and qualified palaeontologist Gunnar Andersson, his conviction that they were significantly older than the bones he had found was to be fully borne out.

Early in January 1903, Nordenskjöld realized the ship might never reach them that season and began laying in seal blubber, because their coal was totally inadequate to see them through the winter, as was their food. To feed themselves they would have to undertake the grisly task of killing and skinning about 400 Adélie penguins at the Seymour Island rookery. Camping nearby with Ekelöf and Jonassen, they began the slaughter on 5 February, Nordenskjöld all the while wishing the ship would arrive so that they could put an end to it.

Little did he realize that the *Antarctic*, fatally holed on 11 January, was a bare 12 miles from them, helplessly locked in the pack-ice, with seven pumps going day and night to keep the water down.

A month later Larsen and his shipmates were afloat on an ice-floe. Their ship having sunk, 20 souls, with two boats and seven tonnes of stores and possessions, were struggling northward towards the coast of Dundee Island and the tiny pimple of Paulet Island. Both were agonizingly distant beyond an obstacle course of grinding pack-ice.

Joined by Andersson before he sailed on 11 April 1902 for South Georgia,

Larsen's winter marine survey and geological programme at the island had yielded the first fossil remains found on that island. But by the time he brought the *Antarctic* back to Port Stanley, the senior zoologist Axel Ohlin had become so seriously ill that he had to return to Sweden, where he died a year later.

They sailed again on 6 September 1902, once more to have her coal replenished at Ushuaia. The Argentine government then paid for the ship's bottom to be scraped and the overhaul of sails and rigging before she left on 5 November, bound for the first proper survey of the coast they had discovered west of Astrolabe Island.

Hindered by abnormally abundant ice, Larsen had found the southern end of the Antarctic Strait blocked by heavy pack-ice. Retreating, he tried to pass east of Joinville Island, but the ship became caught in the pack-ice and was carried 100 miles towards Elephant Island, and then back again.

By that time, 21 December, Andersson, the second-in-command, had decided that the party at Snow Hill would have to be brought to the ship if Larsen could not reach them by 10 February. That meant landing him and two others straight away to go round the mainland coast of Erebus and Terror Gulf and start back with them from Snow Hill if the ship did not arrive in time. After selecting Duse as navigator and one of the Norwegian seamen, Toralf Grunden, who had been ashore with him at South Georgia, the three men, with two months' supplies for nine men, had to be landed at the beach in Hope Bay, which Nordenskjöld had pointed out on the west side of the Antarctic Strait.

Landing there on 29 December, they found that much of the gulf was frozen, so that they had no need to follow the coast. However, they were stopped by open water as they neared Admiralty Sound. Meanwhile, Larsen had worked his way south past Joinville Island by New Year's Day 1903, only to be trapped once more in the pack-ice.

Eleven days later a huge crash had brought everyone on deck as the ice fatally holed the ship, jamming itself under the stern, which was thrust bodily upwards about four feet. The ice had torn away a third of the keel but, as though to compensate, had promptly stemmed the inrush of water.

It was the start of a month of being carried to and fro in the gulf, which ended in the ship's doom in the early hours of 12 February when the ice parted and the pumps could no longer hold the water level down. Larsen had foreseen her fate, and two boats and a mass of stores were on the floe beside her by the time the order rang out to abandon ship.

Only the day before, the three men on the beach at Hope Bay, to which

Hope Bay: X indicates the position of the stone shelter and F indicates the place where the fossils were found

they had returned the day after the ship was holed, had given up hope of the ship's return that season. Realizing that they had only two tents and none too much food with which to survive an Antarctic winter (seals and penguins were curiously scarce), they started to build a stone shelter to move the larger tent into.

The fortunes of the Swedish expedition had reached their lowest ebb. Split into three beleaguered parties, each unaware of the others' fate, the mending of their destiny was to weave a story infinitely more surprising than the harsh reception the Weddell Quadrant had dealt them.

For Andersson, now marooned at Hope Bay, there had been one compensation. On 14 January 1903, the day after their return to the beach, he had gone to look at the shoulder of the big mountain to the south-west of them (later named Mt Flora in honour of his finds) and found an array of marvellously preserved plant fossils that, unlike those on Seymour Island, were immediately recognizable as proofs of a tropical climate in the Antarctic in the Cretaceous period, nowadays dated to 65 to 130 million years ago.

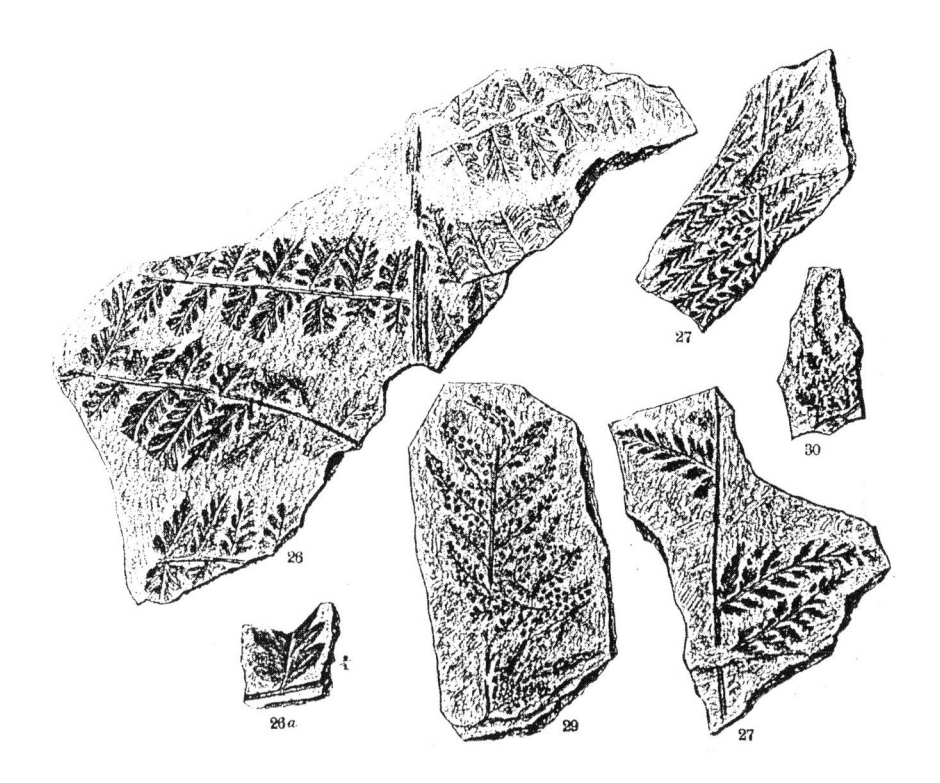

Mesozoic fossil leaves found by Dr Andersson on 1 January 1903
Sphenopteris Nauckhoffiana (no. 26) was the most crucial in dating the plants as having flourished in the Cretaceous period, some 130 million years ago

By 28 February some miracle, and a great deal of hard work by all, had helped Larsen bring the shipwrecked party to the very beach that Nordenskjöld had landed on at Paulet Island.[4]

They were fortunate in arriving before the nearby penguin colonies migrated northward. Whereas Sobral had amassed 33 seals to supplement the Snow Hill party's diet and fuel (the Snow Hill party had ruined three-quarters of their penguin meat, imagining it would be better preserved if they salted it with sea water), the Paulet Island party only found eight seals, so their diet of meat from 1,100 penguins already laid up was only varied with seal on Sundays. All their salt, and much of their sugar, had been lost, along with their mattresses, on the hazardous trip to safety.

Like the others at Hope Bay, they only had a tent to live in, and set about building a stone shelter. With 20 pairs of hands the task was easier. A nearby hill provided a supply of flattish stones for the walls, and long-since deodorized penguin guano made an effective mortar. In a space 20 ft by 18 ft they placed their sleeping bags on stone shelves either side of a 4 ft walkway, beneath a roof of sailcloth held up by tent poles and eventually weighed down with seal skins.

In contrast to Nordenskjöld's confidence about the ship's return, which buoyed him up during the long winter months, Andersson faced a steadily growing conviction that the ship might have been lost, and that the only hope of rescue for his party was to reach Snow Hill. But it was not until 2 September 1903 that the weather allowed him to ski along the old track past Mt Flora to check the state of the Gulf.[5]

Andersson returned with the good news that the way looked safe for sledging. They had their meagre equipment ready on 20 September, after repairing virtually everything with the crudest devices; they had only one sail needle and even unravelled the legs of one pair of ruined socks to darn the gaping holes in the others. With all set to go, an eight-day blizzard drove them back to the shelter until they woke to a clear day on 29 September.

This time they took 12 days to reach the little depot they had left on the far shore on their first attempt. Camped beside it on 10 October, they had seen that there was a channel some 13 miles west of them that curved away southward, but little suspected that there was another tent behind the end of the coast they had now reached.

Fog towards Snow Hill (south-west of them) next morning led to Duse discovering that they were on an island, with Sidney Herbert Sound really being the entrance to the channel that emerged west of them. The more direct routes proving impossible, the three men set off westward on 12 October to sledge through the newly discovered channel.

They halted for lunch about one o'clock, barely three miles to the good. They soon found themselves handing the field glasses to each other in puzzlement at the sight of two seals apparently standing upright on their tails. In a moment the images resolved into the figures of two men. Minutes later they were greeting Nordenskjöld and Jonassen amid the barking of their six dogs. They had to explain who they were before the men from the west could understand, so unrecognizable had the three men's ragged blackened appearance rendered them.

If Duse had discovered an island the day before (they called it Vega Island

Abandon ship! Larsen's dramatic photo of the fatally-holed *Antarctic* as the crew abandoned ship

after Baron Nordenskjöld's famous Northeast Passage ship), Otto Nordenskjöld, the baron's nephew, had just proved that Ross's Mt Haddington stood on a much larger island, separated from the mainland by the channel he had just traversed in a 70-mile pioneer journey from Lockyer Island.

By a strange coincidence they started out on 29 September, the very day the others left Hope Bay. The two men realized only when they had covered the 14 miles to Lockyer Island that they had forgotten to pack the bread, and back they had had to go to make a fresh start on 4 October.

Now here they were, emerged from the channel Nordenskjöld had anticipated (which he named after Crown Prince Gustav of Sweden) after just five and a half days' sledging, interrupted by only two days of bad weather. The two youngest dogs were the yearling survivors of the first Greenland litter, and had pulled behind the four veterans of the southern journey to make some remarkable daily runs.

This time, with more justification, Nordenskjöld had counted on finding seals at both ends of the channel, if channel it proved to be, and started with dog pemmican sufficient for only 20 of the 30 days he thought the journey might take. The gamble had paid off, seals being found as soon as they camped at the first headland beyond Cape Foster, and now again there were plenty on which to gorge themselves.[6]

Bolstered by their feast on the evening of that extraordinary day, the six dogs made light work of the reorganized total load of 772 lb, and the five men made the first-ever traverse of Sidney Herbert Sound to complete 24 miles on 13 October. Then, next day, clear of the sound that had become a channel, Nordenskjöld and his party faced the worst conditions of their journey. With all on skis and two men hauling alongside the dogs, which were up to their hocks in the sticky snow, they struggled for hours at less than two miles an hour, in snow that refused to support even the skis, let alone the sledge runners.

It took two interminable days to cover the 20 miles to Cape Gage, where at last they found a surface swept clear of snow by the wind. Once on it, just three hours sufficed for the 12-mile run to the hut, which they entered two years to the very hour after they left Gothenburg on 16 October 1901.

Nordenskjöld had snatched another prize from the jaws of misfortune and unravelled another fragment in the riddle of the peninsula's coastline. Now it was a case of waiting – even if he had a nagging suspicion of his ship's fate, he was certain that a ship would come.

After three weeks of very restricted outings, 8 November 1903 began in no very unusual way. Bodman and the young Akerlündh were away gathering penguin eggs, when someone saw four men approaching from the north. As everyone hurried to meet them, all were convinced the *Antarctic* had come.

As they neared, Akerlündh hurried forward to meet the group, and in an instant their excitement was turned to dismay. There was indeed a ship, but it was an Argentine navy sloop, the *Uruguay*, and not Larsen's which hadn't ever been heard of since she left Ushuaia!

The Argentine government had assigned the ship for the relief of the missing men the previous April, appointing their naval attaché in London, Lt Julian Irizar, to command her. While the ship was being virtually rebuilt, and enormously strengthened, in Buenos Aires, Irizar had consulted Shackleton, Nansen and others, and returned with a great deal of equipment to await the arrival of the Swedish government's relief ship, the *Fridtjof*, hastily chartered and due to reach Ushuaia before 1 November 1903. It had been agreed that Irizar would sail on that day if she had not arrived, and here he was at his goal a mere seven days later.

The return of the *Uruguay*. Seen here being towed to her berth in Buenos Aires with fore and main topmasts missing, mute witness among the jubilations surrounding the second escape of the rescued men

After much anguished discussion as to whether to return and co-ordinate plans with the Swedish commander, Captain Gulden, or to conduct a search themselves (for none then knew of the fate of the *Antarctic*), Nordenskjöld was left to wrestle with the difficult choice while Jonassen sledged the two Argentine officers back to the *Uruguay* to organize help for moving the collections and other essentials to that ship.

At 10.30 p.m. that same evening, the fate of their shipboard comrades was hanging heavily in the atmosphere in the hut when the dogs began barking. Going outside they saw several figures approaching far out on the ice. Thinking that Irizar might have sent men straight away, it was not until Bodman got outside to find six men standing round the flagstaff that the truth dawned on him. A figure stepped forward from the group, and the astonished magnetician realized it was Larsen!

The story was soon told: their boat journey from Paulet Island via Hope Bay had taken a day longer than Irizar's from Ushuaia. After 22 hours' rowing from the deserted beach, the exhausted party had reached the ice edge at two o'clock that morning.[7] Camping on the ice, unseen by Jonassen and the Argentinians on the way to the ship, they had started their 10-mile walk at 3.00 p.m., and here they were. In an instant the cloud of foreboding was dispersed and fortune had turned 8 November into a day of miracles for the Swedes.

By 4.00 p.m. on 10 November, the nine men at Snow Hill Island were aboard, and 10 hours later, as the *Uruguay* sounded her whistle off the bleak Paulet Island beach, Skottsberg and the 12 men left there with him scrambled out of the stone shelter, rubbing their eyes in disbelief. After setting up a depot of supplies that future expeditions could call on, they erected a cross in memory of Ole Wennersgaard, their comrade who had succumbed to consumption and breathed his last in the smoke-filled shelter on 7 June 1903, the fourth man to give his life in the cause of Antarctic exploration during the multi-national campaign on the threshold of the twentieth century.

Irizar had the ship in Hope Bay by lunchtime the following day to recover the priceless collection of fossils that Andersson had left there, and as the *Uruguay* steamed north out of Antarctic Sound, there cannot have been a man aboard who did not believe their ordeal was over.

Yet within 24 hours they were all in the same predicament as Larsen had been in off King George Island. Forced to heave to in a three-day storm off Cape Melville, the Weddell Quadrant bade them a frightening farewell that left the ship with its two principal topmasts sprung in their crosstrees, and only held up by the backstays. And that with the stormiest seas in the world between them and safety beyond Cape Horn!

Realizing that the masts would be doomed in the next serious gale, Irizar gave the only order possible – to let go the backstays holding the topmasts. To everyone's relief they fell without causing injury or serious damage. Fortunate to have enough coal, the ship steamed into Santa Cruz harbour a week later, to telegraph news of the rescue to an astonished Europe.

That very day, 22 November 1903, the ice broke out of Scotia Bay at Laurie Island and at last released the *Scotia*, William Bruce's ship with the 33 men of the Scottish National Antarctic Expedition who had wintered there, oblivious to the drama being played out 400 miles west of them.

Six hundred miles to the north, the Argentines in Buenos Aires had been awaiting the arrival of Dr Jean-Baptiste Charcot, who had sacrificed the aims of his expedition to offer his help. Unaware of the rescue, he was nearing the mouth of the river Plate in the *Français*.

CHAPTER 3

The French Antarctic Expedition 1903–1905

Jean Charcot aboard the *Français*. Picture by courtesy of his late daughter and her husband M. Robert Allart-Charcot

L'Etat-Major (officers and scientists) of the *Français*: (front, left to right) J. Rey, J-B. Charcot, A. Matha; (back, left to right) P. Pléneau, J. Turquet, E. Gourdon

The French Antarctic Expedition 1903–1905

When Dr Jean-Baptiste Charcot had returned from a summer voyage to Jan Mayen Island in 1902, he was already planning to mount an expedition to Novaya Zemlya. Failing to attract financial support for the sort of ship he needed, he had committed his personal fortune to the building of the *Français*. Barely 105 ft long at the waterline, topsail schooner-rigged, she had two boilers serving a midships engine of no more than 125 hp, which was the most that he could afford.

With the ship's keel laid on 15 January 1903, and plans scarcely formed, news of Nordenskjöld's first discoveries had fired his conviction that France should have been part of the great European campaign being waged in the Antarctic. News of Scott's achievements then arrived to reinforce the enthusiasm of the prestigious committee drawn from the Institut Français, the premier scientific society in Paris, that was considering Charcot's Antarctic plan, which was to go south in the Weddell Sea in the hope of extending Larsen's original discoveries. Barely a month later, news of the disappearance of the *Antarctic* and the preparation of the Swedish relief expedition burst into the headlines. Characteristically, Charcot was the first to propose that the *Français* should join in the search for Nordenskjöld's ship, in conjunction with the *Fridtjof*, which had been chartered by the Swedish government.

The *Français* was launched on 27 June, commissioned by the end of July and sailed from Le Havre on 15 August 1903 to the cheers of enthusiastic crowds, only to be struck by immediate tragedy as the towline broke loose, killing a seaman on the forecastle. Leaving again a few days later to pick up instruments and final supplies at Brest, Charcot finally sailed from there on 31 August, scarcely yet assured of the total finances he needed, despite the handsome support of the national daily, *Le Matin*.

The road to adequate funds had at first been as dispiriting in France as anywhere else. Charcot had almost been in despair at the end of May, with only

FFr20,000 plus the promise of instruments, dynamite and 100 tonnes of coal, firmly assured. Then a chance encounter with the editor of *Le Matin* resulted in the launch of an appeal for FFr150,000. When that brought in only FFr90,000, the editor personally donated the remainder.

Even after the Chamber of Deputies had voted a further FFr90,000 at the urging of President Loubet, total funds amounted to less than half the Scottish National Antarctic Expedition's final cost by the time the *Français* entered Buenos Aires in the last week of November, there to learn of the rescue of the Scandinavians.[1]

This meant that the way was open again for Charcot's own exploration to go ahead, but by now he had lost the only man aboard with Antarctic experience. Charcot had sought and received de Gerlache's support from the start and, after he had changed his objective to the Antarctic, de Gerlache had agreed to serve on the expedition. However, after crossing the Atlantic, de Gerlache pulled out at Pernambuco, Brazil following a difference of opinion over Charcot's plans. With him went two naturalists he had recruited for the expedition.[2]

Advice from Nordenskjöld, Larsen and William Bruce, leader of the Scottish expedition, all in Buenos Aires when Charcot arrived, did much to fill the gap. After inviting them aboard the *Français*, Charcot had soon built up a new plan, based on their discoveries, and promised co-ordination of observations with Bruce's Laurie Island base, which was to be manned by the Argentines under tuition from Robert Mossman for the first year. Charcot published the new plan on 4 December 1903.

The main objectives for the first season were exploration of the west coast of the islands bordering the Gerlache Strait and establishing a winter station as far south as possible, by mid-March 1904 at the latest. If conditions allowed, they would leave news of its whereabouts on Wiencke Island, or on Biscoe's Pitt Island, which de Gerlache thought he had seen at 65° 57' S opposite the 'vaste baie ou détroit' which he and Lecointe had taken for Dallmann's Bismarck Strait.

As to the 1904–05 season, doubtless encouraged by the early timing of Nordenskjöld's Gustav Channel journey, the goal for the spring was dependent on how far south their winter quarters would be. If no further south than the 66th parallel, Charcot would sledge eastward across the Peninsula to the coast that Nordenskjöld had discovered. If his party got into trouble after reaching the Weddell Sea coast, they would leave a message cairn at Mt Christensen and make for Nordenskjöld's depot on Seymour Island. If the winter station was

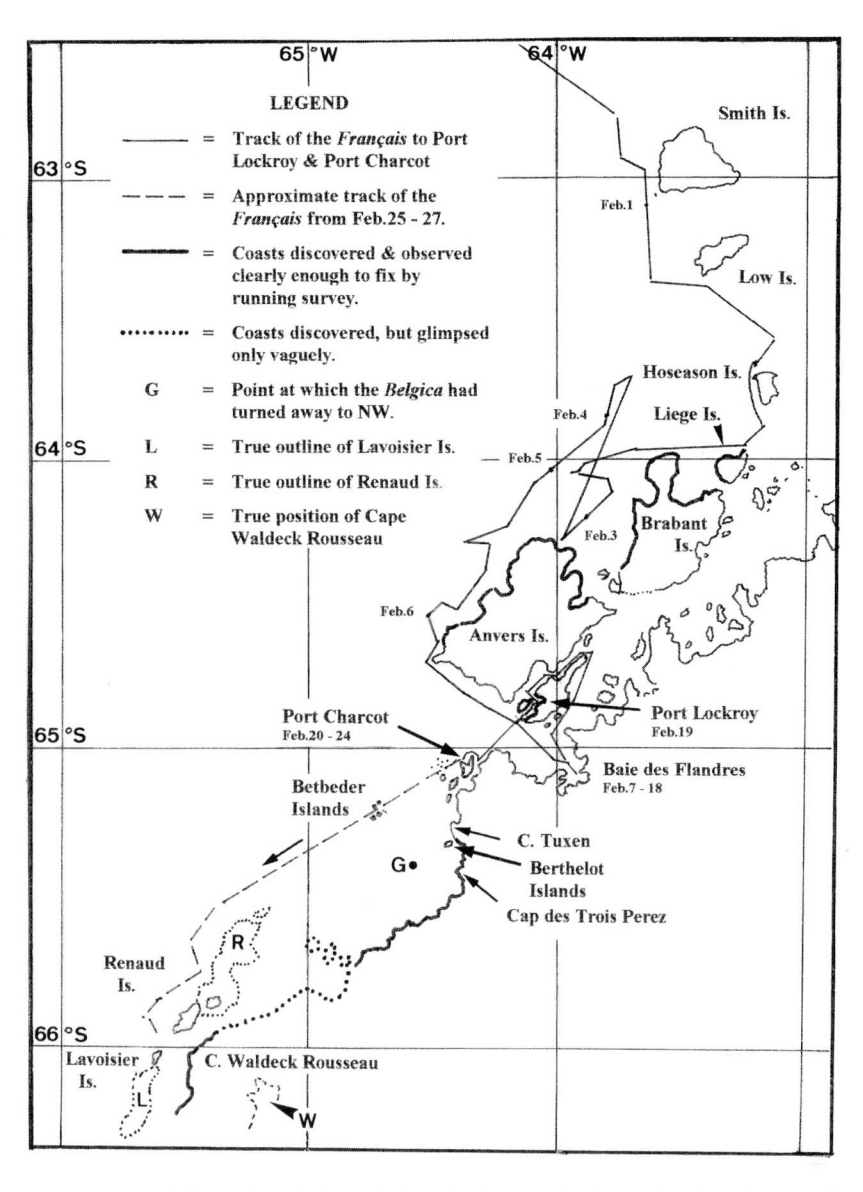

The French Antarctic Expedition 1903–1905: track of the *Français* in 1904, showing coastline discovered; traced from map published with the official expedition journal showing Lt Matha's co-ordinates with longitude converted to degrees west of Greenwich further south, then, more ambitiously still, he would explore the west coast by land as far as Alexander I Land. All that was to be attempted while the ship was still frozen in. Given de Gerlache's pictures of the forbidding nature of the coast in the Lemaire Channel, there was, understandably, no mention of a specific date for the return of the sledge party.

When the ship was freed from the ice they would continue exploration of the coast as far as Alexander I Land, avoiding the fate of the *Belgica* at all costs and returning to South America no later than 1 April 1905. In the event that they discovered a navigable channel through to the Weddell Sea, their course was left unspecified.

The *Français* had arrived in Buenos Aires under tow after a fractured propeller shaft key had stranded her at Montevideo. The Argentine government promised every help with the repair. A government transport would carry their collapsible shore hut to Ushuaia and refill their bunkers there free of charge. The navy would lend him the dogs Nordenskjöld had given to the captain of the *Uruguay*, and the Argentine ship would be sent to look for the cairn the following summer.

With news of the French deputies' vote and contributions pouring in from the expatriate French community, every problem seemed to have been resolved as the *Français* cast off and steamed out on 23 December 1903.[3]

Reaching the Argentine observatory Ano Nuevo on 7 January 1904, they picked up the five best of Nordenskjöld's Greenland dogs, multiplied since left there by Irizar, and steamed westward through the Beagle Channel to anchor at Ushuaia three days later.[4]

With every inch of bunker space crammed with coal, the *Français* was back in the relative shelter of Nassau Bay on evening of 26 January. With the ship watered and swung for compass adjustment, all was ready to cast off from civilization by the following evening, and Charcot and his 19 companions headed south in the smallest ship of all in the great assault on the unknown South that had opened the twentieth century.

They turned west from the entrance to Gerlache Strait six days later and had barely begun their running survey of the seaward coasts of the islands discovered by the Belgians when a tube burst in the port boiler. All its other tubes were found to be blocked by scale. Harassed by fog off the north coast of Anvers Island, and limited to one boiler, Charcot had nevertheless accomplished most of his first objective by the time his ship heaved to at the head of Flanders Bay on 7 February 1904.

Sheltered in an inlet from the stream of icebergs circulating round the main bay, Goudier, the chief engineer, took 11 days to have the boilers repaired and cleaned. De Gerlache had found the bay completely free of ice six years before. With conditions now obviously so much worse, Charcot must have realized that the likelihood of finding a base much further south was slim indeed. Returning north to establish a message cairn on Wiencke Island had become essential, costly as it might be in terms of yet more days lost from the few he now had left.

Sailing north on 19 February, their passage anti-clockwise round the island in search of a suitable site for the cairn had brought them almost full-circle when, on the second day, an inlet near the south-west corner proved to be a channel separating the land west of de Gerlache's Cape Errera from the main bulk of the island. Following the coast eastward into its wide entrance, they came to a perfect little anchorage a mile in from the Neumayer Channel, which Charcot later named Port Lockroy. There was even an ideal site for the cairn on a little island (Casabianca Island) at the entrance of the next inlet to the north, visible from any ship that passed it from either direction. From that moment Charcot knew that he had a safe base if he failed to find one nearer the goal beyond Cape Tuxen that dominated his hopes.

The prize that might be his was tantalizingly near to the other side of the Baie de Flandres lay de Gerlache's Lemaire Channel. That led, after a mere 17 miles, to the towering Cape Tuxen and the vast bay or strait that de Gerlache (and Nordenskjöld on hearing the Belgian's account) had believed was the real Bismarck Strait to the Weddell Sea. He, Charcot, could prove whether it existed, one way or the other.

His hopes of reaching it were soon dashed. The channel was blocked by a jostling mass of pack-ice far too heavy for their ship's puny engine to force them through. As First Lieutenant Matha took the ship round west of the first two islands – de Gerlache's Wandel and Hovgaard Islands – Charcot realized that they were really the Booth and Krogmann Islands in Dallmann's Kaiser Wilhelm I group.

Soon thwarted by innumerable bergs and pack-ice caught up in the maze of small islands outside the main trio, there was little alternative but to retreat to shelter on the north side of Booth Island, where they had been lucky enough to spot a likely-looking cove. Successfully inching the ship into it, astern of Matha, who was sounding from the dinghy, Charcot had at least found a winter harbour a little further south, should all else fail in the coming three weeks.

Charcot's *Français* under sail in the pack-ice and (below) anchored at Port Charcot before wintering there in 1904

Some of the crew beside the hut they erected on shore for emergency stores and shelter in case the ship was crushed by the pack-ice

Another attempt to get through the channel next day proved hopeless after only four miles and forced a return to Français Cove, as Charcot later named it. After watering the ship with melted ice and snow on 23 February, another tube burst as they raised steam the following morning, this time in the starboard boiler. Yet another precious day was lost.

In calm and magnificently clear conditions on 25 February, they were soon among floes so packed that it was impossible to stream the log (to record distance travelled) as they struggled through the cordon of islands and rocks. As they broke out on a south-westerly course, the ice steadily forced them too far west to allow them to see the coast beyond Cape Tuxen. Far away on the horizon lay an isolated island which Charcot took for the most northerly of the Biscoes.[5] Setting course to pass south of it, larger islands appeared to the south, though the view everywhere was fragmented by a patchy mist. Enough could be seen, however, to show that there was no hope of passing inside those islands, and Charcot headed for the open sea.

With calm weather still prevailing, the mist lifted enough the next day to reveal a series of islands, backed by distant snowy peaks on what could only be the mainland. A promising-looking bay on the largest island (Renaud Island) proved to be full of dangerous rocks and icebergs, offering no hope of a landing or harbour.

Beyond the island, another attempt to push towards the land on 27 February soon put them in ever-increasing danger of becoming trapped by the ice. By the time they reached 65° 58' S 66° 22' W further progress had become impossible.[6] The coast, visible beyond the islands, was forbiddingly hostile and obviously bereft of any safe haven for their small ship. It was time to settle for the cove on Booth Island. In considering plans for the British expedition, Markham had surely been right when he dubbed the way south on either side of the peninsula a route of 'little chance'.

As though to emphasize its hostility, the wind rose to a stormy north-east gale, engulfing them in fog and snow showers with little respite as they fought their way northward with rapidly ailing boilers, only to find their cove full of ice and growlers on the night of 3 March 1904.

Much to Charcot's relief, Matha's sortie in the dinghy next morning revealed that they could get safely into the cove, which was too shallow at its entrance to let the icebergs threaten them. It meant they could avoid the retreat to the harbour at Wiencke Island.

That very day the Scots had reaped their reward as they discovered the coast of Coats Land on the eastern shore of the Weddell Sea. For the French there had been only repulse by the coast that had defied both de Gerlache's approach, six years before, and Evensen's five years before that. Now, for the third time, it had hidden its secrets in a swathe of mists behind a barrage of ice.

Charcot, forced to winter no further south than Booth Island, had consoled himself that Français Cove, in the bay he named Port Charcot, lay as far south as Robertson Island and was exactly suited to supplementing the observations by the Swedes at Snow Hill and Mossman's at the Scots' Laurie Island base.

With the *Français* disencumbered once the huts, instruments, stores and dogs were safely installed ashore, and an awning erected over the forecastle, the 20 men of the French expedition soon had their larder augmented with copious rations of seal, penguin and fish. Only two men found it impossible to stomach seal meat, although they readily ate penguin meat instead. Adélie penguins and seals remained on the island, so that by mid-winter they had amassed enough to last them through to the spring.

In spite of the varied menus, on which Argentine beef and veal figured six times weekly, fish three times, seal three times and penguin once a week, Matha came to Charcot on 18 July 1904 to report that he felt too ill to carry on with the multiple duties he had been carrying out – hydrography, surveying, meteorological observations and pendulum gravity measurements amongst others. His rather drawn face, following fainting fits four days before, had hardly perturbed Charcot, for they all looked pallid and tired by that time. But here was his second-in-command with swollen limbs, suffering from palpitations and obviously in the last stages of exhaustion on precisely the calendar day that Hanson, on Borchgrevink's expedition, had developed identical symptoms, before his eventual death in October just five years before.

Remembering with relief that Lecointe's similar symptoms, in July 1898, had been cured in two weeks by Cook's treatment, Charcot had Matha spend many hours each day naked in front of the red-hot stove. His obvious signs of scurvy were hard to credit, for it couldn't be said that their diet had lacked fresh raw meat of one kind or another. All Charcot could do otherwise was treat the cardiac symptoms, augment the patient's diet with copious quantities of fish and tinned milk, and keep him in his bunk.

The fear of losing Matha added a third dimension to the threats already building against Charcot's spring sledging plans. Still centred on resolving the enigma of the strait beyond Cape Tuxen, his every sight of the coast had underlined the impossibility of a land route for exploration. Furthermore, since the previous month the sea ice had also begun to look like a totally unreliable route for a sledge journey. Even in mid-June the southern bay of the island, on their direct route to Krogmann and Petermann Islands (de Gerlache's Hovgaard and Lund Islands), was swept almost clear of ice. If it could break up then, what chance had they of sledging across it in the spring?

Nevertheless, he had to try to establish a depot on one of the two islands. A dangerous, but successful, ski trip with Pléneau, Gourdon and the dogs down the Lemaire Channel to the far island, a week after Matha had reported sick, convinced Charcot that the southern end of Krogmann Island offered the better site. Three days later, with Matha showing signs of steady improvement, he succeeded in placing the first supplies there, the three men accompanied this time by Rallier du Baty, and Paumelle in charge of the two sledges pulled by the dogs.

The journey progressed painfully slowly, with Charcot in front probing every step of the way, and might well have been disastrous, for, after their return that same evening, a storm burst upon the cove. The howling wind and the breaking swell, grinding the ice in the cove against the hull, threatened to end the expedition and smash their small ship to pieces, trapped, as it was, between the shore and the advancing masses they could faintly see at the gates of their haven.

In the gloom of the polar night, amid the threatening din of the storm, they hastily evacuated the ship, carrying Matha ashore in his sleeping bag to lie in the portable hut among the orderly ranks of stores, now invaded by a jumble of vital supplies and equipment from the ship. For eight hours the others lay in their tents outside, the threat of being marooned and losing their ship hanging over them until the storm began to die down.

By 5.00 a.m. it had passed, and Charcot and his weary team returned to their bunks for a few hours' sleep. The ordeal was repeated on 4 August, when, for the second time, the ice massed at the entrance to the cove and the swell threatened to hurl the ship on to the land. Once again, Charcot ordered everyone ashore as the wind suddenly reversed and the perilous prospect of shipwreck loomed. For several hours he felt that their luck had run out and that they would be condemned to more than a year in tents awaiting rescue. But, again, the storm subsided, and they all got back aboard.

For Charcot the relief that they had survived was tempered by the certainty that his sledging plans were in ruins. They would have to make the spring journey by boat, and who was to say they would survive among ice driven by such storms? But at least the cove had saved them, and if the ship could survive those two storms, he could not imagine worse. They would start when the weather improved.

Following two vain attempts to set up a depot on Krogmann Island, during which fog once stranded them on an islet barely 20 yards square, Charcot managed to reach the island only on 3 September 1904.

Intent upon reaching his first goal beyond Cape Tuxen in the whaler, he had at last succeeded in augmenting the small emergency depot sledged to the island on the day of the first storm. This time, his party woke to find themselves marooned there by the state of the ice. Three days later, at considerable risk among the grinding floes, the party regained their base just in time to escape the next in a succession of north-easterly storms that held the ice in their bay. Attempts to establish a second depot on Petermann Island were repeatedly thwarted during the greater part of the following three months.

If that frustrating period saw the worry about Matha lifted when the second-in-command improved sufficiently to resume his surveying at the end of September, Charcot's patience had been tested to the limit when, once more stuck on Krogmann Island for three days, no sign of a break in the ice showed to the south. Climbing the island's highest peak he thought he saw a ribbon of water running south along the mainland shore. They would just have to use that.

Back at the ship, final preparations for the crucial journey began on 21 November 1904 and, when the ice looked feasible in the southern bay on 24 November, Charcot gave the word to go.

With his four companions (Gourdon the geologist, photographer Pléneau, midshipman Raymond Rallier du Baty, a qualified mercantile officer and son of an admiral, and the seaman Besnard) and the boat loaded with 20 days' supplies, a collapsible sledge and the instruments, they set off at two o'clock in the afternoon. They dragged the boat, weighing perhaps 850 kg in all, over the ice for three hours before reaching open water. It was almost dark when they pulled the boat ashore on an islet off the southern tip of Krogmann Island. A second exhausting day brought them to Petermann Island and, on 26 November, they fought their way over to the mainland only to find that the channel leading south was no such thing.

The return to the island was followed by days of interminable labour – 14 hours on 27 November, 18 the next day, and scarcely less to bring the boat to shore at the foot of Cape Tuxen on 29 November. Here, leaving Pléneau in camp suffering badly from snowblindness, the other four scaled the jagged 3,000 ft summit of the cape, there to find that they could not see into the bay they hoped would prove to be a strait. However, some four miles to the south, out among the pack-ice, lay an island from which they could settle the question once and for all.

Leaving a cairn at the foot of the cape, they began the long haul over the uneven ice, often dragging their unwieldy load when up to their waists in half-frozen water. Not until 2.00 a.m. on 1 December could they at last launch the boat in an open channel. Arriving at their goal half an hour later, it took the weary men another three hours to find a landing place adequately sheltered from the north-east gales they had come to fear most.

The next morning Charcot had his reward as he hastened to climb the 650 ft summit of the island, which proved to be the largest of the Berthelot Islands, at 65° 22' S, rather than a single one. The answer to the enigma of Dallmann's strait lay before him – there was not the least sign of a strait. Deeply indented into two bays, the ice-capped coast stretched south to another deep bay beyond the Cap des Trois Perez, as it was later named,[7] and then south-west to Cape Garcia about 30 miles distant.

Favoured with two more days of fine weather, they were able to take angles from the highest of the Argentine Islands (65° 15' S 64° 15' W) on the second day of their return journey, reaching Petermann Island the same evening. Hampered by snow and mist, a long third day saw the five men safely back at the ship, still trapped by the ice filling the bay outside Français Cove.

If 1904 had yielded resolution of the first great question about the coast of Graham Land, denied to de Gerlache from his course a mere four miles further away, how much more might they see on the way to Alexander I Land?

A week later there seemed to be a fair prospect of blasting and sawing a channel to open water 650 yards away along the shore. After five days' work, they woke on 18 December to find Nature had finished the job for them. The bay outside was clear of ice. Work began immediately on recommissioning the engine.

Leaving the steam launch, huts and a depot of supplies with a record of their achievements and intentions in a large cairn, all was ready by Christmas Eve and, after the customary celebration, the *Français* cast off as the tide reached high water.

But it was not to the south-west that they went, for, clear as the bay was, the pack-ice formed an unbroken plain to the horizon across the direct route south. Charcot planned to outflank it by heading out from Dallmann Bay after making a full survey of the harbour at Wiencke Island, which they would aim to touch at on their return.

Arriving there the next day, they completed the task by 30 December, establishing that Mt William was not the highest peak on Anvers Island. A higher peak in the centre of the island was promptly named after their ship. Trapped there for four more days by ice blocking the exits, they weighed anchor early on 4 January 1905 and, stopping only to alter the message in the cairn, emerged into an open sea north of Anvers Island three days later.

Forced far outside de Gerlache's and Evensen's tracks by the pack-ice and the dangers of fog, and by now seriously short of coal for emergencies, Charcot forbade raising steam, except in extremity, as they headed south. Rapidly overtaken by north-easterly storms, and twice narrowly avoiding icebergs in the fog, they saw nothing of the coast until the evening of 11 January, when the horizon suddenly cleared to reveal a conical summit south-east of them. Hasty reference to de Gerlache's charts revealed that it could be nothing other than Alexander I Land.

Matha's sights put them at 67° 25' S, some 40 miles west and 75 miles north of its position according to the Belgians when they had sighted it twice from a westerly course just north of the 68th parallel. But the pack-ice stood across their line of sight as far as the eye could see in either direction, and after 30 hours of probing east and west, Charcot gave up and followed its edge north-east on 13 January, hoping for another opportunity. Twice more that evening they sighted the prominent mountain (later named Mt Bayonne), at about the same time as a snow-capped peak (later named Mt Gaudry) rose over the horizon ahead, with a chain of mountains dwindling away north of it.

Retreating north-eastward that night, with their first sight of other mountain peaks ahead of them, Charcot and his men were some 20 miles south of Biscoe's reputed position when, also approaching from the west, he had first sighted Adelaide Island on 14 February 1832.[8]

When the wind calmed down the following evening, Charcot had no hesitation in ordering steam and, as the ship's speed picked up to six knots, the short four- or five-mile coast, which was all the fog allowed Biscoe to see when he had to turn away a few miles from it, lengthened northward. For the approaching Frenchmen it seemed that this must be the mainland. In the clear weather they were enjoying, there was every possibility that they could follow it north to their furthest south of the previous February.

Even so, the edge of the pack-ice was moving north-west across their direct path, and early on 15 January the French ship had to push through it to reach an open channel, barely two miles wide, along the piedmont ice cliff fringing the coast. As the channel was populated by menacing, moving icebergs, Charcot reasoned that it must therefore be free of submerged rocks and safe for the *Français*'s 10 ft draught. She was well into the channel by 8.00 a.m., at 66° 40' S and some five miles up the coast from Mt Gaudry, when Matha came on the bridge to take over.

Charcot could not bring himself to leave the bridge as they continued at six knots towards a cape, behind which a further coast lay hidden. Then suddenly, without warning, as a tremendous shock literally bent the foremast, the two men saw the forecastle rising up almost vertically before them. Near enough to grab the handle of the engine room telegraph and avoid being flung to the deck, Charcot slammed it round to 'Full Astern' as the ship slipped back into the water and half-clothed men tumbled out of the hatches. Almost mesmerized by the realization of what had happened, Charcot could hardly take his eyes from the water ahead as the deadly rock appeared, glimmering dully beneath the surface.

As his ship reeled back from the submerged rock that had holed her, the starkest of prospects faced Charcot. If they were marooned on the lifeless coast he had discovered there would be no hope of survival. Not a penguin or seal was to be seen on or below the 12 ft piedmont cliff, and they might hardly get more than the barest essentials on to it before the ship sank. If the ship were to sink, that would have to take place in open water where, slim as their chances would be, they might at least be able to reach their base in the boats.

With water rising in the crew's quarters, as the bow slowly began to sink, the crew of the *Français* realized that the pumps had been arranged with no suctions forward of the main bunkers and so were useless until a way could be cut for the water to reach them.

Setting the men to break through the bulkhead and clear a way for the water through the bunkers, now fortunately emptier, Charcot climbed to the main yard and, with Matha and Rey taking turns at the wheel, worked the ship gingerly through the pack-ice to the open water beyond.

Though by that time 45 minutes of pumping had succeeded in lowering the water to such a level that the next shift need not begin for 15 minutes, the situation was scarcely less black; before the accident the condenser had become blocked, rendering the steam drive to the pumps useless. Round-the-clock hand

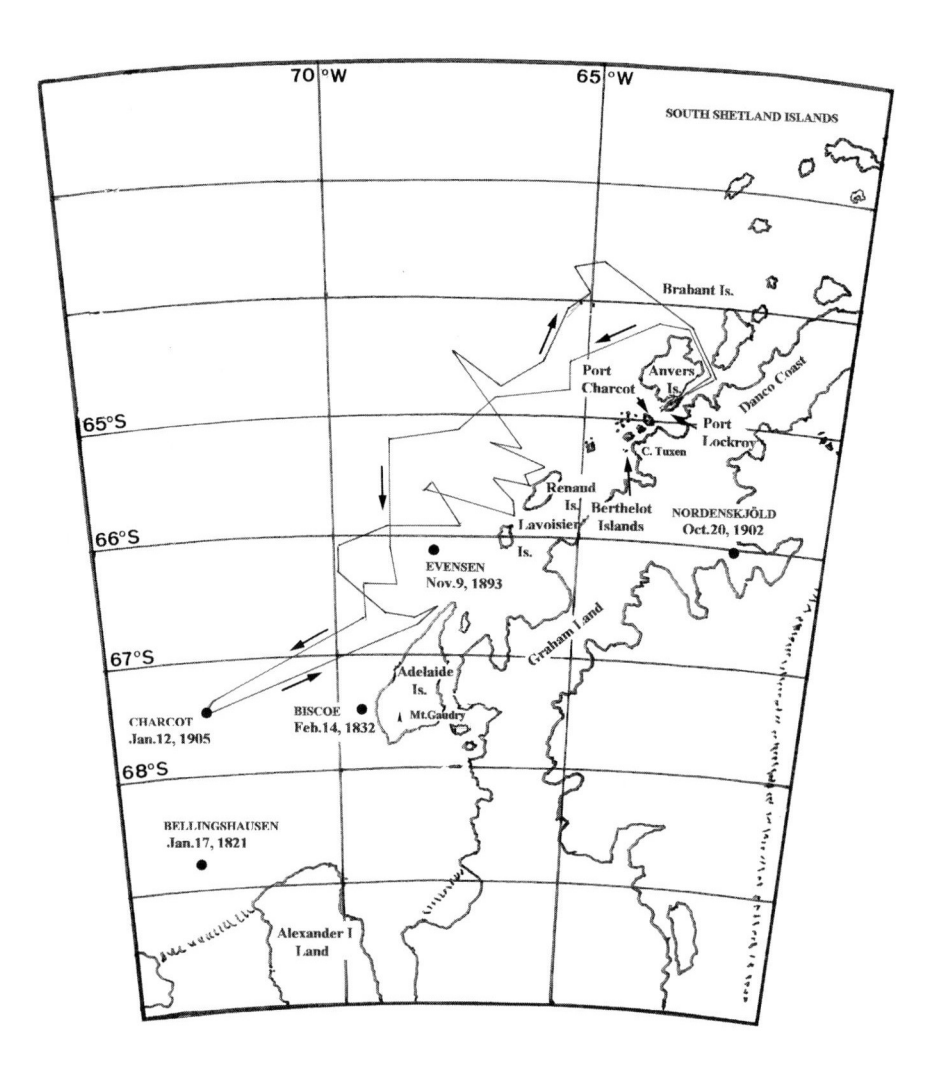

The French Antarctic Expedition 1903–1905: track of the *Français* in January 1905

pumping would be necessary until they reached Ushuaia, or alternatively landed everything and abandoned ship at Deception Island or one of the harbours they had discovered. As the edge of the pack-ice led them away north-west, hours of working by the carpenter in near-freezing water, vainly attempting to caulk the leaks, offered no hope of avoiding that gruelling prospect.

Gratifying as his discovery had been (40 miles of coast that, to him, certainly extended another 20 southward to Mt Gaudry on Biscoe's Adelaide Island, which he had promptly named 'Terre Loubet' after the French President, not realizing it really was an island[9]), the price of escape might yet be beyond their means.

Despite the perilous condition of the *Français*, Charcot, magnetized by the prospect of linking up his new discovery with the mainland at Cape Garcia, could not resist the opportunity to follow a route as near to the coast as the pack-ice would allow. The unknown stretch between 'Terre Loubet's' northern cape and the land they had sighted the previous February was only some 60 miles long. Readily as every man agreed to Charcot's plan, the pack-ice kept that coast hidden, and, when the opportunity came to turn shorewards the following afternoon, they were already back north of the 66th parallel.

Now some 40 miles from their previous closest to the coast, the wind rose to storm force accompanied by blinding walls of snow blotting out all visibility as they hauled round on the port tack. Somehow the ship rode the seas sweeping in on the beam, and survived even when a huge wave broke over the deck, smashing the whaler from its cradle and on to the bridge, where it stove in its planking.

For 12 more hours the ship struggled on until the wind fell calm towards midnight. To their amazement, the exhausted men on the pumps were finding that their task was becoming easier, for by that time little more than 30 minutes in the hour sufficed to hold the water to manageable levels.

Until then at the mercy of any slight accident, with even the cook needed to make up one of the three-man pumping shifts, the good news persuaded Charcot to make one more attempt. That not a murmur of complaint arose when, with steam raised, he headed the bow south-eastward at noon on 18 January 1905 rather than towards safety, must be one of the most extraordinary examples of loyalty in the history of Antarctic exploration. It was no wonder he was later to dub them 'brave gens des gens braves' – bravest of the brave!

The reward was not long in coming, and the penalty not far behind. By midnight three rounded islands were in sight with the mainland beyond them, and soon afterwards, much more clearly this time, the cape they had glimpsed the previous February. When the pack-ice stopped them this time, a couple of miles short of the strait between the two largest islands (Renaud Island and Lavoisier Island), almost 20 miles of the coast was visible. Clear as it was, Charcot was as deluded as to the islands' size as Biscoe had been; Matha's chart shows Renaud Island as barely a quarter of its real extent.

By noon on 20 January, after five hours' work had sufficed to complete Matha's survey, the weather had again closed in, and they headed up the coast of Renaud Island among milling icebergs.

By morning they were once again fighting their way in the teeth of the first of a succession of storms that kept Charcot and Matha on the bridge, taking four-hour watches by turn, for seven long days. When the two men at last manoeuvred the ship into the shelter of the Schollaert Channel, between Brabant and Anvers Islands, at midnight on 28 January 1905 Charcot was able to take his boots off for the first time in 13 days. Headway under sail had been impossible for most of the last week, and, as though to express the engine's exhaustion, they were no sooner in the channel than a key on the propeller shaft sheared.

With that, and the crew's utter fatigue, there was no question of heading north without a respite at Port Lockroy, which they reached that night. 'Oh! Etre au calme, savoir le bateau en sécurité, et pouvoir dormir!' – Charcot's unutterable relief breathes from every word on the page of his expedition journal.

With pumping still necessary for half of every hour, the minimum engine overhaul and the repair of the whaler took 12 days, by which time everyone had recovered. After making a kind of test run to visit the bay beyond Cape Lancaster on the southern point of Anvers Island, where Biscoe had landed in 1832, they headed north on 12 February, with Matha doing a running survey of the mainland coast in the Gerlache Strait as they went.

With Hoseason Island, at the entrance to the strait, astern on 14 February, they had to fight against headwinds nearly all the way (they could not even see the Ano Nuevo lighthouse, let alone put the dogs ashore there), until the *Français* dropped anchor in Puerto Madrin harbour (on the Argentine coast, halfway between the Falklands and the river Plate) on 4 March 1905.

Here they were greeted as if returned from the dead, for, despite there being nearly a month to go to his planned latest return, hope had already been given up following the failure of the *Uruguay* to find the cairn at Casabianca Island. Going on to look for it further to the south, the Argentines had returned empty-handed after being stopped by the pack-ice at 64° 57' S.

For all of the eight days the *Français* lay at Puerto Madrin, the pumps were still manned day and night, and only after the ensuing 10-day voyage to a rapturous reception at Buenos Aires on 22 March could they at last be abandoned as the ship was hurried into dry dock.

Charcot could be well pleased with his achievement. Even if he had only extended certainty about the peninsula's continuity to the point where Nordenskjöld's journey had revealed its other coast, his 'Terre Loubet' was assuredly not the pimple-sized island Biscoe had taken it for when he named it after King William IV's consort, Queen Adelaide of England.

Taken with Larsen's voyage on the Weddell Sea side, 12 years before, and glimpses of the coast south-east of the island, he had taken a very large step towards the mapping of a single land extending to 68° S.

The question of what lay between his 'Terre Loubet' and Alexander I Land was to be as inexorable a lure to the doctor turned explorer as the Pole would be to Shackleton. In the event, the insularity or continental connection of the two lands was not to be finally resolved until more than 30 years had passed, when the British Graham Land Expedition at last revealed the truth.

For now, in 1905, Charcot could rest on his laurels and have a concern for his own fortunes. He had paid for the *Français*, at much financial risk to himself, and knew that a vessel altogether bigger and more powerful would be needed for his next expedition, so when the Argentine government offered to buy the tough little ship, he accepted without hesitation.

Far more suitable than the *Uruguay* for the ice-ridden seas of the Weddell Quadrant, given new boilers and a bigger engine, and renamed *El Austral*, she was put to work the following season on the resupply of the meteorological station at Scotia Bay in the South Orkneys. From there the *Uruguay* had recently brought back the two Scotsmen who had served another year there in the observatory, which the Argentinians would keep up until the Falklands War nearly 80 years later.

The offer to buy the *Français* had come only a few days before they were due to sail, so work on packing the collections for the holds of a commercial liner delayed their departure. Finally, on 5 May, the expedition sailed to the accompaniment of 'La Marseillaise' and cheers from the French cruiser *Dupleix* – the first sign of the official welcome awaiting them in France.

They sailed direct to Tangier, where they found the French government had not been slow to recognize their achievement. They were all transferred to a cruiser, which carried them to Toulon, there to be reunited with their families and welcomed by the mayor. Reaching Paris by train the next day, 10 June 1905, they were greeted by a large delegation headed by the minister of the marine, who, there and then, on the platform at the Gare de Lyon, pinned the Chevalier's badge of the Légion d'Honneur on Charcot's tunic.

At Charcot's request, it was also awarded to all the other officers and scientists. Charcot himself, with Ship's Master Cholet, Chief Engineer Goudier, Leading Seaman Jabet and Midshipman Rallier du Baty, received the Palmes Académiques. The entire crew received the Marine Medal of Honour.

For Charcot, whose wife had divorced him on grounds of desertion, and not a few of the others, it was far from the end of the story. Hardly more than three years later they were back at Buenos Aires in the brand new barque rigged *Pourquoi Pas?* to pursue the quest that haunted Charcot's ambition from the moment he had glimpsed a mountain summit far to the south-east of Mt Gaudry – was Graham Land a peninsula or an island?

Historical Notes and Lists of Expedition Members

Chapter 1: The Belgian Antarctic Expedition 1897–1899

1 Lecointe (*Aux Pays des Manchots*, Brussels, Schepens, 1904, p. 19) refers to four sailors, but de Gerlache (ibid., p. 71) lists five Norwegians, and Lecointe (*Quinze Mois dans l'Antarctique*, Brussels, Bulens, 1902, p. 88) refers to one of these, Adam Tollefsen, as bosun.
2 £1,500 of this would be recovered by sale of the ship.
3 Gaston Dufour and Jean Van Mirlo, the latter assigned as apprentice mechanic.
4 There is some doubt about this, because Lecointe (op. cit., pp. 132–3) relates that the new plan was to return to South America for the winter, then return to the Weddell Sea if they had found good prospects to the south, or, if not, follow the ice to South Victoria Land and carry through the original plan. But is seems unlikely that de Gerlache would have prolonged the first push south by taking soundings across the Drake Strait if he had intended to return that way.
5 Writing over 25 years later, Amundsen claimed that de Gerlache 'forbade any of the ship's company to indulge in' eating seal or penguin, of which Amundsen and Cook had killed 'a great number' (*My Life as an Explorer*, London, Heinemann, 1927, p. 27). When the commandant and Lecointe were bedridden, Amundsen took over command and 'put everyone on seal'. Even de Gerlache ate his share, according to Amundsen, and 'all improved greatly within a week' (this evidently referring to the second outbreak).
6 Roland Huntford in *The Last Place on Earth* (London, Hodder and Stoughton, 1979, p. 73) says that the Belgian Geographical Society had insisted the command remain in Belgian hands. Amundsen's chagrin is understandable, intent as he was on obtaining his master's certificate.
7 Van Mirlo, Dufour, Knudsen and Koren.

Members of the Expedition

Commander	Lt Adrien de Gerlache de Gomery	(31)	Belgian
Navigator and Second-in-command	Lt Georges Lecointe	(28)	Belgian
Surgeon	Dr Frederick A. Cook	(32)	American
Magnetician	Lt Emile Danco	(28)	Belgian
Geologist and Meteorologist	Henryk Arctowski	(26)	Polish
Zoologist	Emile-George Racovitza	(29)	Rumanian
Asst Meteorologist	Antoine Dobrowolski	(25)	Polish

Second Officer	Roald E. G. Amundsen	(26) Norwegian
Third Officer	Jules Mélaerts	(22) Belgian
Chief Engineer	Henri Somers	(34) Belgian
Second Engineer	Max Van Rysselberghe	(19) Belgian
Mechanic and Ordinary Seaman	Jean Van Mirlo	(20) Belgian
Steward and Cook	Louis Michotte	(29) Belgian
Bosun	Adam Tollefsen	(31) Norwegian
Leading Seaman	Ludwig Hjalmar Johansen	(25) Norwegian
Seamen	Engelbret Knudsen	(21) Norwegian
	Auguste Wiencke	(20) Norwegian
	Johan Koren	(18) Norwegian
	Gustave-Gaston Dufour	(21) Belgian

Chapter 2: The Swedish Antarctic Expedition 1901–1903

1 Nordenskjöld in *Antarctica: or Two Years amongst the Ice of the South* (London, Hurst, 1977, p. 130) refers to climbing a small conical mountain above the depot to see the entire coast running west to Cape Foster. That is not possible from Cape Jefford, five miles further west, above the only other practical site for the depot.

2 Nordenskjöld wrote that his pedometer, on average, showed 50,000 paces at the end of each day, and that the daily distance was about 35 km. The instrument counted the paces by the leg it was attached to, so there would have been twice that number of steps. Each one therefore averaged only 35 cm, which is hardly surprising considering the load was not less than 75 lb per man.

3 The inlet de Gerlache believed might be the real Bismarck Strait was at 65° 25' S. Sobral's position fix for the depot camp (65° 48' S 62° 11' W) suggests their advance in the teeth of the wind to the Borchgrevink Nunatak, on 18 October, constituted some 15 miles in six hours. Despite the lighter load, that would have been altogether remarkable, given their crossing of what must have been the Leppard Glacier tongue. Nordenskjöld does not record the mileage on 11 October, the first day's march from the Castor Island nunatak (easternmost of the so-called Seal Islands), but, subtracting his estimate of 57 miles for the subsequent advance to the depot camp from Sobral's position for it, implies only 12 miles on that day, with the wind directly aiding them.

It is altogether more likely that the mileages were the other way round – for example, 15 on 11 October and 12 on 18 October. That would place the depot camp at 65° 51' S, south of the Flask Glacier tongue, the position adopted for this narrative. Such a position is also suggested by Nordenskjöld's reference to soon encountering 'unpromising ice' that persuaded them they would 'fall in with a new glacier' the day they started north from the depot on 22 October. There is no reference to such a surface in his account of the journey south towards it, which is consistent with their approaching from the north-east having avoided its outermost crevasses.

Outward- and homeward-bound weather conditions clearly prevented them seeing either glacier, let alone the 30 miles up the Leppard Glacier to its summit, or Nordenskjöld would have understood the nature of the Richthofen Valley with its seemingly flat surface curving into the latter, about 10 miles distant from him, as he photographed it from the Borchgrevink Nunatak.

4 See list of officers and crew below for those in the Paulet Island party.
5 The peninsula they had traversed to reach the gulf is today named Tabarin Peninsula, after the World War II operation led by the naturalist James Marr, who had been with Shackleton in the *Quest* in 1921, and then, after a distinguished polar career with Mawson aboard *Discovery* in 1929–30, and with the Discovery Investigations aboard *Discovery II*, was chosen to set up a base in Hope Bay in 1944 to prevent it being used by enemy submarines and raiders.
6 Nordenskjöld (op. cit., p. 307) implies that they had camped at Red Island, but the implication that they had covered 15 miles from there to Cape Well Met by 1.00 p.m., having already had their lunch break, is hardly credible, let alone consistent with his description of a 'short southward march' and of 'unbroken ice' in the direction of Paulet Island. The author's conclusion is that they camped on Corry Island. Although it is hard to equate it with Nordenskjöld's description of it as 'in the middle of the channel', his reference to passing 'a narrow deep bay with Mt Haddington in all its magnificence in the background' (this would be the western end of Sidney Herbert Sound with Croft Bay behind it) during the afternoon is hardly consistent with the camp being anywhere west of that island. The day's run to Corry Island would have been 25 miles. After daily runs of 15 and 17 miles through the channel, the longer distance ties up well with his description of the run on 11 October as 'a good long day's march'.
7 Larsen's companions were Karl Andersson, Second Engineer Georg Karlsen, Third Mate Axel Reinholdz, Bosun Anton Olsen Ula and Cook Axel Andersson.

Members of the Expedition Staff
Snow Hill Shore Party

Leader and Geologist	Otto Nordenskjöld	(32)
Hydrographer and Meteorologist	Gösta Bodman	(26)
Doctor	Eric Ekelöf	(26)
Scientific Assistant	Lt José Sobral	(?)

Hope Bay Party

Second-in-command and Palaeontologist	John Gunnar Andersson	(?)
Cartographer	Lt Samuel A. Duse	(27)

Aboard *Antarctic*

Zoologist	Axel Ohlin	(34)	Invalided after first voyage
Asst Zoologist	Karl Andreas Andersson	(26)	In Paulet Island party
Botanist	Carl Skottsberg	(21)	In Paulet Island party
Artist	Fredrick Wilbert Stokes	(43)	First voyage only

Officers and Crew of the *Antarctic*

Captain	Carl Anton Larsen	(N)	(41)	B/S, PI
First Mate	F. L. Andreassen	(N)	(43)	B/S, PI
Second Mate	H. J. Haslum	(N)	(45)	B/S, PI
Third Mate	Axel R. Reinholdz	(S)	(28)	B/S, PI
Chief Engineer	Anders Karlsen	(N)	(37)	B/S, PI
Second Engineer	Georg Karlsen	(N)	(18)	B/S, PI

Blacksmith	Ole Johnsen Bjönerud	(N)	(33)	B/S, PI
Carpenter	(unknown; hired for first voyage only)			B/S
Sailmaker	Ole Olaussen	(N)	(21)	B/S, PI
Steward	G. F. Schönbäch	(S)	(22)	B/S, PI
Cook	Axel Andersson	(S)	(25)	B/S, PI
Bosun	Anton Olsen Ula	(N)	(40)	B/S, PI
Able Seamen	Ole Jonassen	(S)	(27)	B/S, SH
	Gustav Akerlündh	(S)	(18)	B/S, SH
	Toralf Grunden	(S)	(27)	B/S, HB
	Martin Tofte	(N)	(50)	B/S, PI
	Ole P. Duus	(N)	(21)	B/S, PI
	Ole C. Wennersgaard (died on Paulet Island)	(N)	(20)	B/S, PI
	J. Aitken	(FI)	(22)	PI
	F. Jennes	(E)	(22)	PI
Firemen	Karl Johanssen	(S)	(?)	B/S, PI
	Wilhelm Holmberg	(S)	(?)	B/S, PI

Note: S = Swedish, N = Norwegian, FI = Falkland Islander, E = English, B/S = Buenos Aires–Snow Hill (first voyage) PI = Paulet Island Party (second voyage), SH = Snow Hill Party, HB = Hope Bay Party

Chapter 3: The French Antarctic Expedition 1903–1905
1 The final total before leaving Buenos Aires amounted to FFr450,000, which, by the labour and supplies they provided, the Argentines augmented to an extent that virtually eliminated the threat of insolvency.
2 Charcot gives this explanation in the expedition journal published after the expedition, but his biographer, Mlle Marthe Emmanuel, wrote that de Gerlache pleaded unhappiness at separation from his fiancée, hinting at an unwillingness to play second string to Charcot. The reality may lie in the likelihood that involvement in the search for Nordenskjöld's ship would so limit the scope of the expedition, despite the funds voted by the Deputies, as to forfeit the Belgian's interest, which had until then outweighed his personal feelings. Charcot, who would not have a word said against de Gerlache, would certainly not have revealed the personal side of his companion's motives for pulling out when preparing a document for general publication.
3 In the dark they passed the *Scotia*, stranded on a shoal on her way back to collect Bruce, without realizing the mishap that had befallen her.
4 In *Le Français au Pole Sud* (Paris, Flammarion, 1906, pp. 85–9), Charcot describes only three of the dogs (Sogen, Nerven and the bitch Fia) as pure-bred, the other two being Storm and another bitch Peridota. Storm was not used on the only sledge journey on which the others were employed (the journey to the Krogmann Island depot on 28 July 1904).
5 The 'island' on the horizon was almost certainly the group at 65° 15' S 65° 05' W that Charcot named after the Argentine Admiral Betbeder and, according to the course plotted on Lecointe's chart, the same as de Gerlache's 'Iles Cruls', towards which the *Belgica* was forced away north-west on 13 February 1898, when near the reef today

known as Grim Rock and too far out to see the coast in the 'vast bay or strait'. Today, islands further north bear the name Iles Cruls.

6 Given in the expedition Meteo Report (Matha, A. and Rey, J-J., *Hydrographie, Physique du Globe*, Paris, Musée de la Marine, 1911, pp. 264–8) as the 27 February 1904 position. That was a remarkably accurate fix, compared to the disproportionate size and position assigned to the islands on Lecointe's chart, which was yet another instance illustrating the difficulty of running surveys in misty conditions.

7 The cape Lecointe showed on his chart at 65° 35' S.

8 Hugh Robert Mill, in *The Siege of the South Pole* (London, Alton Rivers, 1905), describes Biscoe as 'running east-southeast to 67° S 72° W where he sighted land on Feb. 14th [1832]', and that he later sounded '3 miles off shore'. The map in Mill's book shows Biscoe's course running on to ± 67° 15' S 69° 15' W, which, incidentally, is very near to de Gerlache's course at that point. If there was reason to doubt Biscoe's chronometer and longitudes from his assertion that he could see the summit of 'Adelaide Island' from 72° W, Charcot's first sighting of the 8,335 ft Mt Gaudry, from about the same distance (± 60 miles) as Biscoe, lends much credibility to their accuracy.

9 Charcot would establish that on his 1908–10 expedition in the *Pourquoi Pas?*

Members of the French Antarctic Expedition 1903–1905

Chef de l'Expédition	Dr Jean-Baptiste Charcot
Geologist and Glaciologist	E. Gourdon
Zoologist and Botanist	J. Turquet
Asst Observer and Photographer	P. Pléneau
Ship's Master (*Patron*)	E. Cholet
First Officer (Hydrography and Gravitation)	Lt A. Matha
Second Officer (Magnetism and Meteorology)	Ensign Jean-Jacques Rey
Midshipman (Cadet)	Raymond Rallier du Baty
Chief Engineer	E. Goudier
Second Engineer/Carpenter	F. Libois
Steward	Robert Paumelle
Cook	M. Rozo
Bosun	J. Jabet
Seamen	A. Besnard
	J. Guéguen
	F. Hervéou
	F. Rolland
Alpine Guide and Acting Seaman	Pierre Dayné
Stoker	F. Guéguen

THE SCOTT POLAR RESEARCH INSTITUTE

The Scott Polar Research Institute in Cambridge, UK, was founded in 1920 as the memorial to Captain Robert Falcon Scott and his four colleagues who died in 1912 on return from the South Pole during the ill-fated *Terra Nova* Expedition. It is the oldest international centre for polar research within a university and houses the worlds' premier Polar library, which holds an unequalled collection of both published and unpublished polar material. There is also a small museum open to the general public. A programme of entertaining public lectures are held on a regular basis.

An association of Friends was established in 1946 to provide individuals with a means by which they could support the important work of the Institute, and to keep members in touch with the Institute and other polar activities. Over the years the Friends have provided valuable funding to the Institute, providing assistance for the Institute's scientific programmes, the acquisition and preservations of archive materials and support for the Library. News and information is circulated through a newsletter and social events are regularly held, giving individuals the opportunity to meet with other Friends and members of the Institute.

To join the Friends of the Scott Polar Research institute, go to their website:

www.spri.cam.ac.uk/friends

or write to The Membership Secretary, Friends of the Scott Polar Research Institute, Lensfield Road, Cambridge CB2 1ER, United Kingdom.

THE UNITED KINGDOM ANTARCTIC HERITAGE TRUST

The Antarctic Heritage Trust was founded in recognition of Britain's long and distinguished role in Antarctic exploration and scientific research. The New Zealand Antarctic Heritage Trust oversees conservation of the huts from the heroic age of exploration in the Ross Sea. Its sister organization, the United Kingdom Antarctic Heritage Trust supports the work in the Ross Sea. In addition it watches over the conservation of heritage sites on the Antarctic Peninsular and South Georgia. Further objectives include promoting an educational programme for young people, the acquisition and preservation of Antarctic memorabilia in the UK, and running the Friends of Antarctica organization.

Members are kept informed of Antarctic news and developments through a newsletter. There is also the occasional opportunity to meet other Friends of Antarctica, usually including a talk or similar entertainment. For full information go to their website:

www.heritage-antarctica.org